CONTINUOUS-FLOW FAST ATOM BOMBARDMENT MASS SPECTROMETRY

CONTINUOUS-FLOW FAST ATOM BOMBARDMENT MASS SPECTROMETRY

Edited by
R.M. Caprioli

JOHN WILEY & SONS
Chichester · New York · Brisbane · Toronto · Singapore

Copyright © 1990 by John Wiley & Sons Ltd.
Baffins Lane, Chichester
West Sussex PO19 1UD, England

All rights reserved

No part of this book may be reproduced by any means,
or transmitted, or translated into a machine language
without the written permission of the publisher.

Other Wiley Editorial Offices

John Wiley & Sons, Inc., 605 Third Avenue,
New York, NY 10158-0012, USA

Jacaranda Wiley Ltd, G.P.O. Box 859, Brisbane,
Queensland 4001, Australia

John Wiley & Sons (Canada) Ltd, 22 Worcester Road,
Rexdale, Ontario M9W 1L1, Canada

John Wiley & Sons (SEA) Pte Ltd, 37 Jalan Premimpin 05-04,
Block B, Union Industrial Building, Singapore 2057

Library of Congress Cataloging-in-Publication Data:
Continuous-flow fast atom bombardment mass spectrometry / edited by
R.M. Caprioli.
  p.   cm.
  Includes bibliographical references (p.
  ISBN 0 471 92863 1 (cloth)
  1. Mass spectrometry. I. Caprioli, R.M.
QD96.M3C65   1990
543'.0873—dc20   90-12683
                                                            CIP

British Library Cataloguing in Publication Data:
Continuous-flow fast atom bombardment mass spectrometry.
  1. Biochemistry. Mass spectrometry
  I. Caprioli, R.M.
  574.19285

  ISBN 0 471 92863 1

Printed in Great Britain by Biddles Ltd, Guildford.

CONTENTS

Contributors   vii
Preface   ix
Acknowledgements   xi

1. Design and Operation   1
    1.1 Design and Operational Parameters   2
    1.2 Stable Operation   4
    1.3 Operational Advantages   11
    1.4 Operational Disadvantages   25
    1.5 Conclusion   26

2. Trace Analysis   29
    2.1 Instrumentation and Methods   31
    2.2 Optimization of Experimental Parameters   33
    2.3 CF-FAB/Tandem Mass Spectrometry of Leucine–Enkephalin   38
    2.4 Determination of Platelet Activating Factor (PAF) in Cell Extract   40
    2.5 Conclusion   42

3. Quantitative Analysis   45
    3.1 Instrumentation   45
    3.2 Parameters Affecting Quantitation   48
    3.3 Focused Cesium Gun   56
    3.4 Conclusion   61

4. Direct Analysis of Biological Processes   63
    4.1 Flow Injection Analysis   64
    4.2 Batch Sample Processing   68
    4.3 On-line Reaction Monitoring   73
    4.4 Microdialysis   80
    4.5 Conclusion   91

5. Liquid Chromatography/Mass Spectrometry   93
    5.1 Microbore/MS Applications   95
    5.2 Capillary (Nanoscale) LC/MS   107
    5.3 Conclusion   118

6. Capillary Zone Electrophoresis/Mass Spectrometry 121
   6.1 General Principles 121
   6.2 Capillary Zone Electrophoresis/Mass Spectrometry 125
   6.3 Conclusion 135

7. Analysis of Low-polarity Substances 137
   7.1 Probe Design 137
   7.2 Analysis of Acetogenins 140
   7.3 Analysis of Metabolites of Benzo[a]pyrene 143
   7.4 Other Non-aqueous Applications 147
   7.5 Conclusion 148

8. Other Applications 151
   8.1 Development of CF-FAB Mass Spectrometry for the Biopharmaceutical Industry 151
   8.2 Biomedical Applications of Gradient Capillary LC/MS Using CF-FAB 160
   8.3 Microbial Biotransformation and Microcolumn LC/CF-FAB MS for Sulfonyurea Herbicide Metabolite Identification 166
   8.4 Analysis of Metabolites of Fatty Acid Oxidation 175
   8.5 Applications of Coaxial CF-FAB with Tandem Mass Spectrometry 181

Index 187

CONTRIBUTORS

Richard M. Caprioli — Department of Biochemistry and Molecular Biology and the Analytical Chemistry Center, The University of Texas Medical School at Houston, Houston, TX 77030, U.S.A.

Simon J. Gaskell — Center for Experimental Therapeutics, Baylor College of Medicine, Houston, TX 77030, U.S.A.

Sanford P. Markey — Section on Analytical Biochemistry, National Institutes of Health, Bethesda, MD 20892, U.S.A.

M. Arthur Moseley — Laboratory of Molecular Biophysics, National Institute of Environmental Health Sciences, Research Triangle Park, NC 27709, U.S.A.

Ralph S. Orkiszewski — Center for Experimental Therapeutics, Baylor College of Medicine, Houston, TX 77030, U.S.A.

David L. Smith — School of Pharmacy, Purdue University, West Lafayette, IN 47907, U.S.A.

Kenneth Tomer — Laboratory of Molecular Biophysics, National Institute of Environmental Health Sciences, Research Triangle Park, NC 27709, U.S.A.

WITH SPECIAL APPLICATIONS BY:

Bradley L. Ackerman — Merrell Dow Research Institute, Cincinnati, OH 45215, U.S.A.

James H. Bourell — Department of Protein Chemistry, Genentech Inc., South San Francisco, CA 94080, U.S.A.

Teng-Man Chen — Merrell Dow Research Institute, Cincinnati, OH 45215, U.S.A.

John E. Coutant — Merrell Dow Research Institute, Cincinnati, OH 45215, U.S.A.

Leesa J. Deterding — Laboratory of Molecular Biophysics, National Institute of Environmental Health Sciences, Research Triangle Park, NC 27709, U.S.A.

John Frenz	*Department of Medicinal and Analytical Chemistry, Genentech Inc., South San Francisco, CA 94080, U.S.A.*
William S. Hancock	*Department of Medicinal and Analytical Chemistry, Genentech Inc., South San Francisco, CA 94080, U.S.A.*
William J. Henzel	*Department of Protein Chemistry, Genentech Inc., 460 Point South San Francisco, CA 94080, U.S.A.*
David S. Millington	*Division of Genetics and Metabolism, Department of Pediatrics, Duke University Medical Center, Durham, NC 27710, U.S.A.*
Daniel L. Norwood	*Division of Genetics and Metabolism, Department of Pediatrics, Duke University Medical Center, Durham, NC 27710, U.S.A.*
Robert W. Reiser	*E. I. du Pont de Nemours & Co., Agricultural Products Department, Experimental Station, Wilmington, DE 19880-0402, U.S.A.*
Ming-Chuen Shih	*Section on Analytical Biochemistry, National Institutes of Health, Bethesda, MD 20892, U.S.A.*
Barry Stieglitz	*E. I. du Pont de Nemours & Co., Agricultural Products Department, Experimental Station, Wilmington, DE 19880-0402, U.S.A.*
John T. Stults	*Department of Protein Chemistry, Genentech Inc., South San Francisco, CA 94080, U.S.A.*

PREFACE

Mass Spectrometry has undergone remarkable advances in the past decade which have literally revolutionized its capabilities and sphere of application in modern analytical technology. Perhaps it has had its greatest impact in the field of biological applications where compounds formerly believed to be intractable by mass spectrometry, because of their charged nature and/or high molecular weight, can now be routinely analyzed. Fast atom bombardment (FAB) mass spectrometry has played a major role in this regard and has proved extremely useful because it can be used to analyze ionic and polar molecules without the need for chemical derivatization. Its success is also partially the result of its ease of use and the fact that it can usually be inexpensively retrofitted on most modern mass spectrometers.

Continuous-flow FAB was born out of the need to further improve the applicability of the FAB technique by allowing the direct analysis of aqueous based solutions. It is clear, insofar as biological applications are concerned, that since a major portion of biochemical processes occur in aqueous solutions, new analytical techniques must be able to deal with this medium directly. Thus, the mass spectrometer can be used as a molecularly specific analytical device to explore the intricacies of an on-going chemical process without perturbing the chemical dynamics of that process. This also opens up possibilities for on-line monitoring of these processes in real time. Further, it was found that the continuous-flow FAB technique also could be operated with mixtures of aqueous and organic solvents and pure organic solvents as well. Because of this and its on-line capability, a number of investigators found that this liquid introduction system had value as an interface between the mass spectrometer and chromatographic, electrophoretic, and microdialysis devices.

This book has been written for the purpose of disseminating information about continuous-flow FAB mass spectrometry, including operational aspects, applications, and specific experiences of a number of investigators in applying it to specific analytical tasks. The book is aimed at providing the novice with sufficient details and operational tips so that the mechanics of the technique can be more quickly mastered, while still providing the experienced investigator with additional information which might allow enhanced performance for specific applications.

For the most part, this book is based on a workshop entitled "Continuous-flow FAB Mass Spectrometry", sponsored by The American Society for Mass Spectrometry, held in November, 1989, in Annapolis, Maryland. Although not a verbatim transcript of the workshop, the chapters of this book are based on presentations made at this meeting. I am grateful to the American Society for Mass Spectrometry for their encouragement in the production of this book.

Richard M. Caprioli
June 1, 1990

ACKNOWLEDGEMENT

The editor wishes to express his thanks and gratitude to Alice Stafford and Margaret Winkler for preparing the copy for this book. Their perseverance and attention to detail is greatly appreciated.

Chapter 1

DESIGN AND OPERATION

Richard M. Caprioli

Desorption ionization techniques have had a considerable impact on the field of mass spectrometry during the past decade (1-4). The further development and application of these techniques in areas involving biological research has truly been phenomenal. Fast atom bombardment mass spectrometry (FAB MS) is a desorption ionization method used widely today because it allows the investigator to obtain mass spectra from polar and charged compounds without the use of derivatization techniques (5,6). In addition, it is easy to use and can be fitted on most modern mass spectrometers.

Basically, the FAB ionization technique utilizes an atom such as xenon, which has been given approximately 5-8 keV of translational energy, for the bombardment of a liquid sample on a target inside the source of the mass spectrometer. The impact of this particle on the surface of the liquid produces what can be envisaged as a molecular sputtering of the surface layers of the sample. A variety of processes take place which lead to the formation of ions in the gas phase (7) and these ions are then analyzed in the spectrometer. It is important to note that the FAB process is basically a surface analysis technique for liquid samples, analyzing molecules which reside in the molecular layers at the liquid/vacuum interface. However, the advantage of the liquid nature of the sample is that the surface is continually renewed by molecules from deeper within the sample droplet.

In order to keep the sample a liquid inside the high-vacuum environment of the mass spectrometer, a viscous organic liquid is added to the sample in relatively high concentrations. Perhaps the most commonly used liquid for this matrix role is glycerol. Typically, a sample is prepared for a standard FAB analysis by taking 1 μl of an aqueous solution of the compound to be analyzed and mixing it with 2 to 4 μl of glycerol. After insertion of the sample probe into the mass spectrometer through the pre-pumping chamber and vacuum lock, the liquid sample is 90% or more glycerol. An extremely wide variety of molecules can be analyzed by mass spectrometry in this manner, including many that were formerly considered intractable by mass spectrometry. Thus, intact and underivatized peptides, sugars, antibiotics, organophosphates, nucleotides, drug metabolites, and many other charged and polar compounds have been successfully analyzed.

The use of the viscous organic matrix liquid in FAB MS also has several significant drawbacks. First, it gives rise to a very intense background chemical noise in the mass spectrum as a result of radiation damage to molecules in the sample at the point of impact of the xenon atoms. Also, for glycerol, intense ions derived from clusters of glycerol molecules can be seen throughout the mass spectrum at m/z values of $(92n+1)$ in the positive ion mode and $(92n-1)$ in the negative ion mode, where n is a positive integer. Second, the matrix liquid essentially contaminates the sample, diluting it and producing poor sensitivity. Thus, high-sensitivity analyses (at or below the picomole level) are extremely difficult in the low mass range below approximately m/z 400. Third, the viscous matrix enhances the ion suppression effect, i.e., the phenomenon in which a compound present in a sample will be recorded at an unusually low intensity or not at all. This has been found, in large part, to be the result of the tendency of hydrophilic compounds to migrate to the interior of a sample droplet, away from the liquid/vacuum interface. At the same time, hydrophobic compounds tend to migrate to the surface layers of the droplet, suppressing the ionization of other compounds (8,9).

1.1 DESIGN AND OPERATIONAL PARAMETERS

Continuous-flow FAB (CF-FAB) was devised to take advantage of the analytical capabilities offered by FAB while at the same time eliminating or diminishing the negative aspects of the technique (10-12). Essentially, the method utilizes a carrier solution, typically 95% water and 5% glycerol, which is allowed to continuously flow onto the target of the probe inside the ion source of the mass spectrometer where the solution is subsequently bombarded. Samples can be superimposed onto this carrier solution by flow-injection or can be contained as a component of this solution. A similar technique, termed Frit-FAB (13), utilizes a fine stainless steel mesh frit at the target to disperse the solution. This device has been used as an LC/MS interface and its design and operation will be discussed in more detail in Chapter 5.

A schematic diagram of the first CF-FAB probe, built for the Kratos MS 50, is shown in Figure 1.1(a). A syringe pump is used to feed carrier solution through a microinjector valve and then through a fused silica capillary inside the probe shaft, terminating at the target or sample stage. Individual samples can be conveniently loaded and injected into the carrier liquid flow through the use of the microinjector. For instruments with high-voltage sources, an insulator is used to isolate the target from the probe shaft. The high-vacuum seal is made by a Swagelock fitting using a teflon or vespel ferrule. The

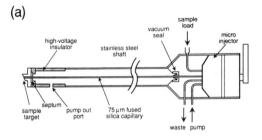

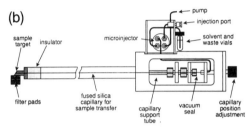

Figure 1.1. Continuous-flow FAB probe designs for (a) Kratos MS 50 and (b) Finnigan-MAT 90 high-performance mass spectrometer. See text for more detail.

septum located directly behind the sample stage should fit snugly around the fused silica capillary in order to prevent liquid on the sample stage from running back into the probe shaft. The tip of the probe comes into contact with the source block and is heated. Since 5 to 10 μl of water are being evaporated per minute, some heating is necessary to prevent freezing of the liquid on the surface of the target or in the exit portion of the capillary. Typically, the source block is maintained at a temperature of about 45-50°C for this purpose. Stable operation of the device is achieved when the rate of flow of liquid onto the surface is balanced by the rate of evaporation. Best performance is achieved when the liquid sample is maintained as a thin film on the target surface.

A second CF-FAB probe design for a high-performance mass spectrometer is shown in Figure 1.1(b) for use with a Finnigan MAT 90 instrument. Although basically similar to the other design, two important additional features should be noted. First, this probe is equipped with a capillary position adjustment device which allows the capillary to be drawn in or out of the plane of the surface of the target while the probe is in operation in the mass spectrometer. This has proven to be a convenient and effective manner to stabilize the liquid flow quickly and achieve maximum performance. In addition, directly under the target, two absorbent filter pads are placed to help remove the excess liquid from the target by capillary action. The pads facilitate achieving a thin film on the target surface and minimize the time needed to reach stable operating conditions.

The size of the fused silica capillary used is not critical, although it obviously affects the flow dynamics. Capillaries having large inner diameters (between 50 and 100 μm) and outer diameters from 100 to 300 μm are most often used. For example, a 1 meter length of 75 μm (i.d.) fused silica capillary used for delivery of approximately 10 μl of carrier solvent per minute will produce only a very low back-pressure at the pump (less than 10 psi). Of course, use of LC columns, trapping columns, additional microvalves and fittings, or other flow restrictors will significantly increase the pressure required to maintain this flow.

The materials used to make the target surface do not appear to be critical. Targets made of copper and stainless steel are commonly utilized, although the former tends to dissolve even if low concentrations of acids are used in the solvent. Stainless steel targets appear to be best overall because they can be highly polished and are easily cleaned by dipping them in a concentrated HCl solution for a few minutes. Excessive cleaning, however, tends to pit the surface and impedes the smooth flow of liquid. The shape of the target similarly does not appear to be a critical parameter; some manufacturers and users prefer flat tips while others prefer rounded ones. While most commercial tips are flat, a round tip is standard on the CF-FAB probe supplied by VG Analytical for the ZAB-high field instrument.

1.2 STABLE OPERATION

There are several criteria which can be used to evaluate operational stability of the CF-FAB probe. These criteria may differ, depending on the analytical task to be performed; that required for qualitative analysis can be different than that needed for quantitative analysis. These are discussed in detail in Section 1.2.4. A number of different parameters have been found to be important in affecting the stability of the device and a brief discussion of these follow.

1.2.1 Backflow

It was briefly mentioned above that the septum placed directly behind the target tip inside the probe is important to prevent capillary action from causing the liquid on the target to backflow into the shaft via the space between the capillary and the hole in the target as shown in Figure 1.2. Once liquid gets into this area inside the target tip, it can vaporize and erupt outward, considerably affecting the surface of the sample and causing sample "spitting".

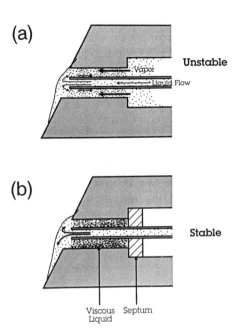

Figure 1.2. CF-FAB target tip showing (a) unstable condition caused by backflow of liquid into the probe shaft and consequent vapor formation and (b) stable condition using a septum to prevent backflow.

Use of a flow stop such as a septum prevents backflow. Thus, the hole in the sample stage should ideally be only slightly larger than the outer diameter of the fused silica capillary so that the capillary can be moved and replaced easily while at the same time a large dead volume is avoided. Any small dead-volume space is filled with glycerol or other viscous liquid matrix at high concentration because the water component readily evaporates. This process, which normally takes 5-10 minutes, is partly responsible for the initial instability on insertion of the probe. When this process is complete, the carrier solution then flows onto the surface and over the viscous liquid matrix seal around the capillary. Of course, several minutes are also needed to equilibrate the temperature of the probe tip to that of the source block. Figure 1.3 shows the initial ion current produced on insertion of a probe at room temperature (time 0). After approximately 5 minutes, the ion current is sufficiently stable to acquire qualitative data and, after about 10 minutes, quantitative data can be obtained.

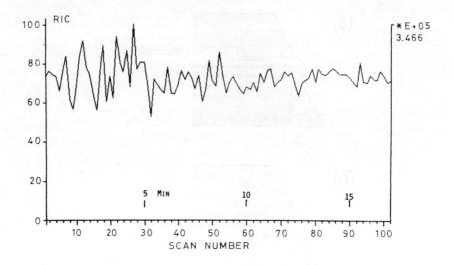

Figure 1.3. Total ion chromatogram recorded on insertion of the CF-FAB probe ($t=0$) showing increasing stability obtained over a 15 minute period.

1.2.2 Achieving a Thin Film

Perhaps the most important criterion for obtaining optimal performance with CF-FAB is achieving and maintaining a thin film on the target surface, as shown in Figure 1.4. Both stable operation and the highest sensitivity occurs when a thin liquid sample film is bombarded. In this regard, we have found the presence of the absorbent pad to be most effective in drawing liquid off the surface by capillary action. At a flow rate of 5 μl/min with a 5% glycerol carrier solution, the pads will last 5-6 hours or more before they need to be changed. In contrast, when a liquid droplet is allowed to form, unstable operation occurs with disruptive sputtering of the liquid and, consequently, poor performance is observed. Liquid sample build-up to form a droplet on the probe tip can result from a dirty target surface where the surface tension is high or by pumping liquid onto the target surface faster than it can be removed by either evaporation or by deposition into the filter pad.

The design of the CF-FAB probe for the Finnigan MAT TSQ 70 is different from those described above in that it utilizes the topical tangential delivery of liquid onto a solid surface. However, the optimum operating condition also involves achieving a thin sample film, depicted in Figure 1.5. Although this design has the advantage that liquid backflow is not possible because the target is solid, it has the disadvantage that the device is not a "probe" that can

Design and Operation

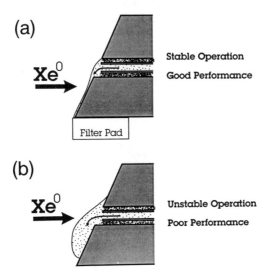

Figure 1.4. CF-FAB target-tip showing (a) stable operation with thin film condition, and (b) unstable operation with liquid droplet build-up.

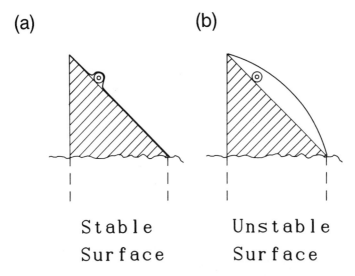

Figure 1.5. Visual appearance of (a) stable and (b) unstable operating condition for the CF-FAB setup for the Finnigan-MAT TSQ 70 with the surface "tangential" delivery design.

be readily inserted and removed. We have constructed a CF-FAB probe for the TSQ 70 (14), shown in Figure 1.6, that is coaxial in design, similar to those

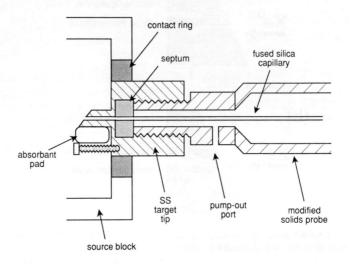

Figure 1.6. Design of the CF-FAB probe for the TSQ 70 with the liquid delivery capillary inside the probe shaft (coaxial design), similar to those shown in Figure 1.1.

described earlier. This probe also depends on contact with the source block as a means of heating the target and utilizes an absorbent pad to help draw off excess liquid from the target.

The maintenance of a thin film on the target surface is therefore the result of the interaction of several dynamic parameters; the temperature of the probe tip, the pumping capacity of the instrument, and the rate of flow of liquid onto the surface. Since the pumping capacity of commercial instruments is fixed, stability is basically a function of temperature and flow rate, as shown in Figure 1.7. Such observations are made using the source block temperature because it is easily monitored; the actual target temperature cannot be conveniently followed on most commercial instruments. Generally, stable operation occurs at a flow rate of approximately 5 μl per minute (range: 2 to 15 μl/min) at a temperature of approximately 45 °C on the source block (range: 40-80 °C). In this figure, the carrier solution is assumed to be 95% water and 5% glycerol. Conditions, of course, will change depending upon the nature of the carrier solution being supplied to the probe tip. With respect to increasing the pumping capacity in the source, a cold trap directly located in the source chamber will give approximately a five-fold decrease in pressure.

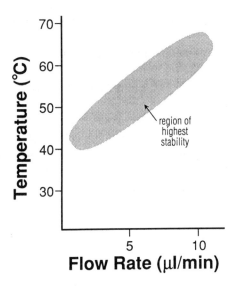

Figure 1.7. General relationship of source temperature and flow rate in providing a stable operating condition for CF-FAB.

1.2.3 FAB Gun Stability

Under certain circumstances, the high-pressure in the source, due to solvent evaporation, can cause unstable operation of the FAB gun itself. This is readily seen in fluctuations of the high-voltage and ion current meters of the gun power supply. This condition can be remedied by replacing the outer aluminum cathode disc with one fabricated with a shaft about 1 cm long which fits inside the gun barrel. Thus, it has the appearance of a cylindrical "top hat" having a hole down the center with the rim taking the place of the original cathode. The exit hole should also be slightly smaller in diameter than that of the standard cathode so that, overall, the conductance of gas back into the gun assembly is significantly reduced.

1.2.4 Achieving Stable Operation

Generally, one can define stable operation as the achievement of the following conditions:

(a)   ±10% or less variation in the intensity of an ion derived from the carrier solvent, e.g., *m/z* 185 (from glycerol), *m/z* 42 (from acetonitrile, if present),

(b) smooth injection profiles (e.g., see Figure 1.10),
(c) reproducibility in the measurement of the area of peak profiles of ±10% or less, and
(d) maintenance of a steady source pressure, e.g., for the MAT 90 equipped with a cold trap, the ionization gauge does not waiver and reads about 2×10^{-4} torr at a flow rate of 5 µl/min.

Absolute values of these parameters will differ somewhat depending on the instrument and operating conditions. Of course, for qualitative analyses, these stability criteria can be relaxed, but for extremely accurate quantitative analyses, they should be strongly adhered to. Figure 1.8 demonstrates the stability obtained with CF-FAB on the MAT 90 for monitoring two ions in the carrier solvent over a period of approximately 40 minutes immediately following probe insertion and a 10-minute stabilizing period. The ion at m/z 185 is the

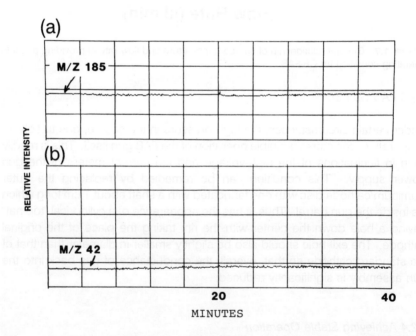

Figure 1.8. Selected ion monitoring of compounds in the carrier solvent (92% water, 5% glycerol, 3% acetonitrile) over a period of 40 minutes with the MAT 90 CF-FAB probe. The ion at m/z 185 is the protonated glycerol dimer and m/z 42 protonated acetonitrile molecular ion.

protonated glycerol dimer and the ion at m/z 42, the (M+H)⁺ ion of acetonitrile, the latter added to this carrier solution at 3% by volume. Similarly, Figure 1.9 shows the stability of the new CF-FAB probe for the TSQ-70 (see Figure 1.6)

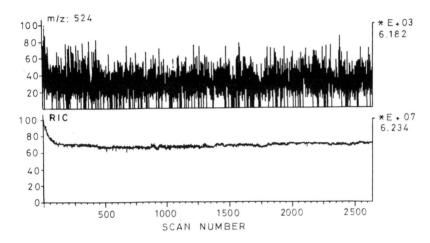

Figure 1.9. Total ion chromatogram *(bottom)* and selected ion chromatogram of m/z 524 *(top)* from monitoring the carrier solvent (92% water, 5% glycerol, 3% acetonitrile) using the coaxial probe design (Figure 1.6) with the TSQ 70.

over a period of several hours. The probe was inserted at scan 1 and scans 1-50 show the initial equilibration period. The instrument was scanned from m/z 40-600. The reconstructed ion chromatogram of a single mass, m/z 524, is also shown in the figure to illustrate the level of the background present. The ion intensities are expressed at the right-hand portion of the figure and, in this case, the background level at m/z 524 is approximately 1/20,000 of the total ion current. The high total ion current results from the low start mass (m/z 40), with major ion intensity derived from glycerol at m/z 93, 185, 277, etc. and fragments of these clusters. Thus, although CF-FAB significantly reduces background relative to standard FAB, a relatively intense background level still remains below m/z 200.

1.3 OPERATIONAL ADVANTAGES

There are several advantages obtained from use of CF-FAB relative to the standard FAB technique. These include:

(a) the capability of direct introduction of aqueous sample solutions,
(b) high relative sensitivity,
(c) decreased intensity of background and matrix cluster ions,
(d) decreased ion suppression effect,
(e) quantitative comparison of sample injections,
(f) high rate of sample throughput,
(g) temporal relationship of sample ions to background ion intensity.

1.3.1 Flow Injection Analysis

A major advantage of CF-FAB for the analysis of ionic compounds lies in the ability to inject small amounts of an aqueous solution directly into the mass spectrometer via the carrier solution (10-12). This process confers several advantages; it eliminates extraction procedures for the purpose of eliminating water, it concentrates the sample on the target tip over a relatively small period of time (30 to 60 seconds) giving maximum signal intensities, and provides a temporal relationship between the sample ions and background ions derived from the carrier solvent. The net result is the production of a "chromatographic" peak, permitting the effective subtraction of background mass spectra from the mass spectra of the peak itself. Figure 1.10 illustrates

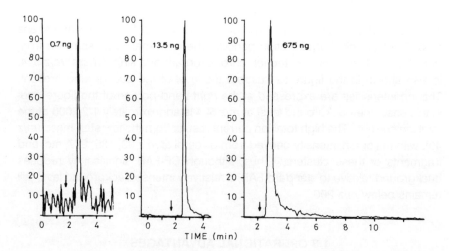

Figure 1.10. Selected ion chromatograms of the molecular ion region (m/z 1345-1351) from injections of the indicated amounts of substance P in 0.5 µl water. The arrows indicate the points of injection. (Reprinted with permission from reference 10.)

the flow injection profiles of an aqueous sample of the peptide substance P (MW 1347 daltons) obtained from the injection of 0.5 µl of each of three solutions of different concentrations. Peak tailing, the phenomenon that leads to memory effects and limits the rapidity of injections, is significant only at the higher concentration. Figure 1.11 shows the linearity of the response in terms

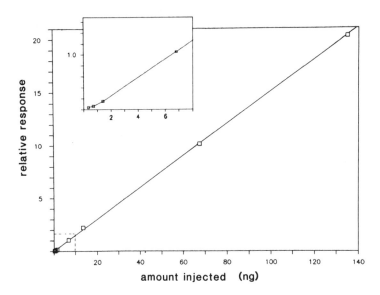

Figure 1.11. Linearity of response for the range 0.35-135 ng of substance P injected in 0.5 µl water. The response was calculated from the area of the peak in the selected ion chromatogram from triplicate injections. The average deviation from the mean is less than ±10%.

of the areas calculated under the injected peak profiles for a series of samples ranging from 0.35 to 135 ng injected. Each plotted data point is the average of three separate injections.

Although quantitative aspects are discussed later, a brief comment here is appropriate. When the criteria for stable operation are met, as described above, one is able to relate the areas of the peak profiles obtained from consecutive sample injections with a reproducibility of ±5-10%. For example, ten replicate injections of 1 µl of a solution of the peptide substance P at a concentration of 200 pmol/µl gave a peak area reproducibility of ±7% (standard deviation). Use of stable isotopically labeled analogs and standards can be used to significantly improve the reproducibility beyond this level and is discussed in more detail in Chapter 3.

1.3.2 Sensitivity

The sensitivity of CF-FAB can be as much as 200-fold greater than that provided by standard FAB. Although this occurs primarily as a result of a decrease in background chemical noise and, therefore, actually represents a lower limit of detection, it is also partially the result of increased ion production from the sample (11). For simplicity, the result of the combination of these effects will be referred to as an increase in sensitivity. Figure 1.12 shows a

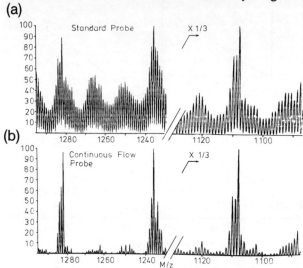

Figure 1.12 Portion of mass spectra taken with (a) standard FAB and (b) CF-FAB from a sample containing angiotensin II (mw 1045) and III (mw 930). This mass region is above the $(M+H)^+$ ions and shows the background obtained with the two methods. The observed peaks are copper adduct ions of the molecular species. (Reprinted with permission from reference 10.)

portion of the mass spectrum of a peptide for a mass range above the $(M+H)^+$ ion obtained with standard FAB and CF-FAB. This region of the spectrum shows the copper adduct ions formed as the result of the interaction of acid in the sample with the copper target used in this case. It is seen quite clearly that a substantial reduction in the background chemical noise has occurred in the mass spectrum obtained with CF-FAB. Sufficient detail can be seen in the spectrum to allow the copper isotope pattern to be observed and the number of copper atoms in the complex to be calculated. This same effect is also shown in Figure 1.13 for a wider mass range for the peptide des-glutamine substance P. A substantial decrease in the background chemical noise is seen in the spectrum obtained with CF-FAB to the extent that ions derived from fragmentation of the peptide, particularly those for loss of side

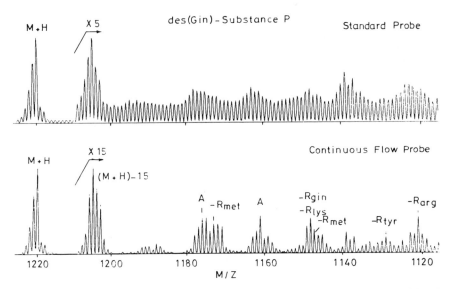

Figure 1.13. Comparison of the FAB spectra taken with (a) standard FAB and (b) CF-FAB for 10 pmol of the peptide des(gln)-substance P. Loss of side chain groups from the molecular species and the position of type A cleavage fragment ions are indicated. (Reprinted with permission from reference 11.)

chain moieties, can be seen quite clearly. These ions are obscured by the background in the spectrum obtained with standard FAB. In addition, it is interesting to note that the peak 16 mass units below the molecular ion, attributed to radiation damage of the molecule caused by the bombardment process, is decreased in intensity by a factor of 3 with CF-FAB. This effect is not unique and has been observed in all the mass spectra of peptides. It is believed that this is simply the result of the greater aqueous nature of the sample being bombarded. Figure 1.14 shows the mass spectrum obtained with CF-FAB for the injection of approximately 500 pmol of the oligosaccharide maltoheptaose. Again, background chemical noise and cluster ions produced from the glycerol matrix are significantly reduced.

The sensitivity increase is most apparent when small amounts of sample are injected. Figure 1.15 shows this effect for the injection of 5 pmol and 100 fmol of the peptide substance P. At the 5 pmol level, the sensitivity enhancement using CF-FAB, relative to that of standard FAB, is approximately 30- to 50-fold. At the 100-fmol level, the molecular ion obtained with standard FAB is not discernible above the background chemical noise, while that obtained with CF-FAB gives a signal-to-noise ratio of approximately 4-5 to 1. This represents a sensitivity increase of approximately 150-fold for this sample.

16   CF-FAB Mass Spectrometry

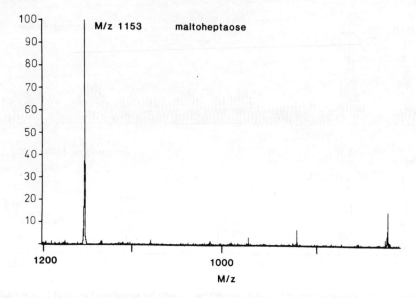

Figure 1.14. Mass spectrum of 500 pmol maltoheptaose between m/z 800 and 1200 taken with CF-FAB.

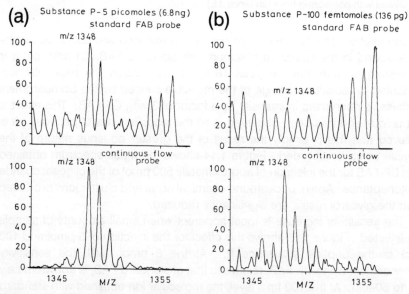

Figure 1.15. Comparison of FAB mass spectra taken with standard FAB using 90% glycerol/10% water and CF-FAB using 80% water/20% glycerol as the solvent. (a) injection of 5 pmol, and (b) 100 fmol of the substance P. (Reprinted with permission from reference 11.)

It is noted that in the standard FAB spectrum, m/z 1348 is recorded at about 4000 counts, while that in the CF-FAB spectrum is about 400 counts, i.e., in the former, 90% of the signal is derived from background. Figure 1.16 shows the

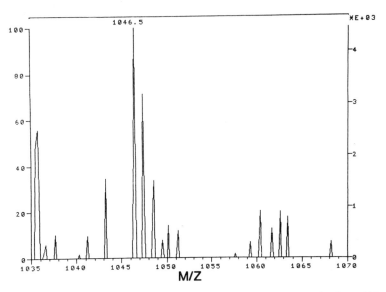

Figure 1.16. Background subtracted spectrum of the molecular ion region from 10 fmol of angiotensin II (mw 1045.5 daltons) taken with CF-FAB.

background subtracted spectrum of the molecular ion region for the analysis of 10 fmol of the same peptide, approximately the limit detection for this compound (a scan window of 5 mass units used). Finally, the overall effect of the decreased background over the full spectrum is shown in Figure 1.17 for the analysis of 200 pmol of substance P. The standard FAB spectrum shows a low-intensity molecular ion species with the intense ions at lower mass being derived from protonated glycerol clusters. The spectrum obtained with CF-FAB shows a rather intense molecular species, a reduced chemical background, and greatly decreased intensities for the glycerol cluster ions. The molecular ion species is increased 55-fold with CF-FAB over that obtained with standard FAB at the 200 pmol sample level.

1.3.3 Decreased Ion Suppression

Another significant advantage of the use of CF-FAB is the tendency of the technique to diminish the ion suppression effect (9, 15). This effect is characterized by the low intensity or absence of a specific ion if the compound

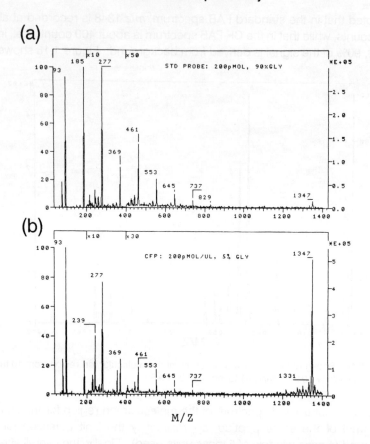

Figure 1.17. Comparison of the spectra of 200 pmol of substance P taken with (a) standard FAB and (b) CF-FAB. The molecular ion species intensity obtained with CF-FAB is approximately 55-fold greater than that achieved with standard FAB.

tends to migrate to the interior or more hydrophilic portion of the sample droplet rather than to the surface of the droplet where bombardment and gaseous ion formation actually take place (2). In mixtures one can also observe that hydrophobic compounds, which prefer to reside in the surface layers, will suppress the ion intensities of more hydrophilic compounds. The benefit of CF-FAB in this case is illustrated in Figure 1.18 for a portion of the mass spectrum of the tryptic hydrolysis of myoglobin. The $(M+H)^+$ ion for one fragment at m/z 1515 is totally suppressed in the spectrum obtained by standard FAB while a significant peak can be recorded using CF-FAB. It is believed that this decrease in the ion suppression effect is simply the result of the thin layer formed on the target and the mechanical mixing which occurs as the liquid flows over the target surface. However, the ion suppression effect

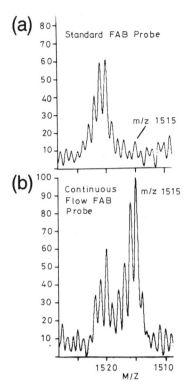

Figure 1.18. Partial FAB mass spectrum of the tryptic digest of myoglobin. The ion at m/z 1515 represents residues 119-133 of the protein. Spectra were obtained with (a) standard FAB and (b) CF-FAB for 100 pmol of the digest protein. (Reprinted with permission from reference 9.)

is not completely eliminated using CF-FAB; extremely hydrophilic compounds, especially in the lower mass range, can still exhibit significant suppression.

A second example of the benefit of CF-FAB in minimizing ion suppression is shown in Figure 1.19 for the analysis of a mixture of seven different synthetic heptapeptides listed in Table 1.1 (15). With standard FAB, peptides labeled I, III, IV and V are significantly suppressed compared to the spectrum shown in the bottom panel obtained with CF-FAB. These three heptapeptides are multiply charged in the acidic medium used in this analysis (approximately pH 1) and are very hydrophilic. Compounds labeled II, VI and VII in Figure 1.19 are singly charged peptides at this pH and show the same relative intensities in both spectra. It is also noted that the glycerol cluster ions are absent or significantly reduced in the spectrum obtained with CF-FAB.

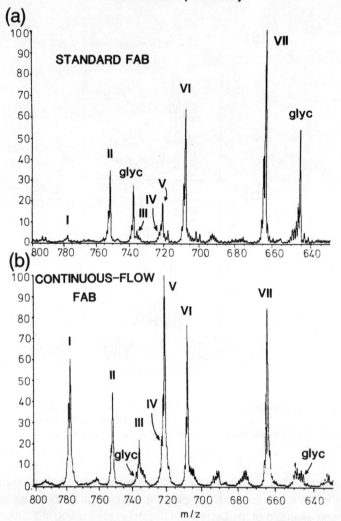

Figure 1.19. Analysis of a mixture of seven peptides with (a) standard FAB and (b) CF-FAB. The peptides are listed in Table 1.1. Ions derived from glycerol are marked "glyc". (Reprinted with permission from reference 15.)

To further characterize this phenomenon, Table 1.2 shows the comparison of the ion intensities for standard FAB and CF-FAB for a number of peptides produced in the tryptic digest of myoglobin. In almost all cases, ion intensities obtained with CF-FAB are greater than that obtained with standard FAB. At lower masses, ion suppression occurs with CF-FAB as well. Even though the hydrophilicity index for the peptide at $(M+H)^+$ 708 is less than that at m/z 1515, the smaller peptide is much more hydrophilic than this calculation

Table 1.1. Comparison of Ion Suppression Effects for CF-FAB and Std-FAB with Hydrophilic Peptides

Compound	Sequence	$[M+H]^+$	Hydrophilic index[a]	Charge (at pH 1)	Relative ion intensity[b]	
					Std-FAB	CF-FAB
I	Ala-Phe-Lys-Lys-Ile-Asn-Gly	777.4	37	+3	4	59
II	Ala-Phe-Asp-Asp-Ile-Asn-Gly	751.3	80	+1	35	64
III	Ala-Phe-Lys-Ala-Lys-Asn-Gly	735.4	331	+3	7	31
IV	Ala-Phe-Lys-Ala-Asp-Asn-Gly	722.3	353	+2	4	18
V	Ala-Phe-Lys-Ala-Ile-Asn-Gly	720.4	59	+2	28	100
VI	Ala-Phe-Asp-Ala-Ile-Asn-Gly	707.3	80	+1	70	100
VII	Ala-Phe-Ala-Ala-Ile-Asn-Gly	663.3	80	+1	100	100

[a]Calculated by the method of Naylor et al. (8) from the data of Bull and Breese (18).
[b]Measured from an average of five mass spectra, with reproducibility of repetitive analyses less than ±10%.

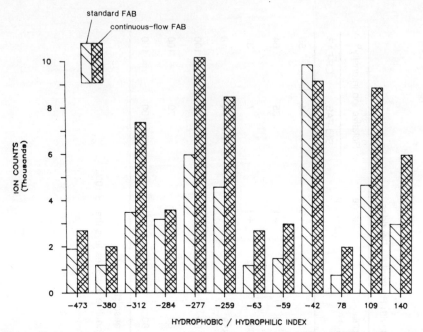

Figure 1.20. Plot of the hydrophobic/hydrophilic index of tryptic fragments of bovine β-lactoglobulin A versus recorded ion counts for standard FAB and CF-FAB. The masses of the (M+H)$^+$ ions for the peptides, from left to right are m/z 1066, 1194, 917, 2313, 838, 904, 2031, 1659, 673, 1637, 934, and 1246. (Reprinted with permission from reference 15.)

Table 1.2. Analysis of the Tryptic Digest of Myoglobin

[M+H]$^+$ of peptide	Hydrophilic index[b]	Relative ion intensity[a]	
		Std-FAB	CF-FAB
1927	-425	33	32
1894	-288	5	8
1855	+285	-	13
1593	+58	67	76
1515	+428	-	33
1392	-238	100	100
950	+132	-	83
748	-502	736	1647
708	+140	-	-

[a] The intensity of m/z 1392 was arbitrarily taken as 100. A dash denotes a signal intensity not distinguishable above background.
[b] Derived from data from reference 18.

predicts. As peptides get smaller, the hydrophilic contribution of the N-terminal amino group and C-terminal carboxyl group increases in proportion, a factor not included in calculations of hydrophilicity in Table 1.2. In a second study of the suppression effect in complex peptide mixtures, the comparison of CF-FAB with standard FAB was made for peptides in the tryptic digest of β-lactoglobulin. Figure 1.20 shows the results with the peptides ranked according to their hydrophobic/hydrophilic index. Again, one can see that ion intensities obtained with CF-FAB are generally significantly greater than those obtained with standard FAB.

Together, the reduction in the background noise and the diminished ion suppression effect can have a significant effect on the spectrum of a given compound. This is illustrated in Figure 1.21 for the analysis of 500 femtomoles

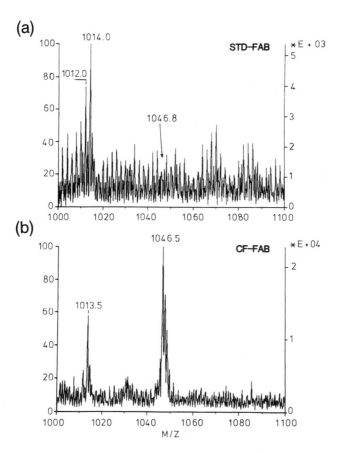

Figure 1.21. Mass spectra of 500 fmol of angiotensin II obtained with (a) standard FAB and (b) CF-FAB.

of the peptide angiotensin II. For the spectrum obtained with standard FAB, the $(M+H)^+$ ion of the peptide, which should appear at m/z 1046, is not discernible above the background chemical noise. The same amount of sample produced a spectrum with CF-FAB which shows an intense molecular ion at m/z 1046. Another example is taken from the analysis of a synthetically prepared peptide which was being checked for purity following cleavage of the synthetic material from the resin. The peptide is quite hydrophilic and, using standard FAB, the spectrum could not be obtained because of the presence of glycerol and background chemical noise (Figure 1.22(a)). The sample was analyzed with CF-FAB and this spectrum is shown in Figure 1.22(b). This

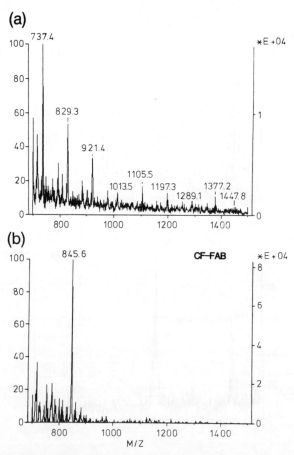

Figure 1.22. Analysis of a hydrophilic synthetic peptide with (a) standard FAB and (b) CF-FAB. The latter spectrum was able to be background subtracted (see text). (Reprinted with permission from reference 12.)

analysis also illustrates the advantage of the temporal factor introduced by flow injection analysis using CF-FAB because it allowed background to be efficiently subtracted. The net result is an intense molecular species at m/z 845.6, the ion expected for the putative peptide synthesized.

1.3.4 High Mass Analysis

The use of CF-FAB does not appear to provide any greater capability for high mass analysis than that obtained with standard FAB. The molecular species centered at approximately mass 5735 for bovine insulin gave quite similar spectra under comparative conditions in both cases, although different matrixes were used. Standard FAB requires the use of thioglycerol to obtain a good signal-to-noise ratio for the molecular species of this compound, while for CF-FAB, the inclusion of either 5% glycerol or 5% thioglycerol gave nearly identical spectra. This phenomenon has proven to be generally true for the analysis of peptides at both high and low mass, that is, the importance of particular matrix compounds seems to be diminished with CF-FAB. However, this may not be true for other types of molecules. Several reports have shown that other matrix compounds added at the low levels are also helpful in producing superior spectra with CF-FAB (16, 17).

1.4 OPERATIONAL DISADVANTAGES

The use of CF-FAB is accompanied by some specific operating disadvantages. Certainly one of these is that the operation of the instrument is somewhat more complex in that stable operation involves control of a dynamic process in which the balance of liquid delivery and evaporation must be maintained for optimal performance. Operating pressures inside the ion source are higher than that with standard FAB so that for high-voltage sources, voltage instability may be encountered. Another disadvantage is that an initial stability period of 10 to 15 minutes is generally required before satisfactory performance is achieved. Also, great care must be taken to eliminate dead volumes in the capillary connections used in the system. These dead volumes can lead to significant peak tailing and will limit the rapidity of the injection process. Generally, we have found that the injection of 0.5 μl of a solution of a compound at a carrier flow rate of 5 μl/min will lead to a sample peak of approximately 30-50 seconds at baseline. Therefore, maximum sample injection throughput is approximately one sample every 2 minutes. Another drawback is the fact that large volumes of water and some glycerol are being evaporated and/or sputtered in the ion source. Particularly in the case of quadrupole instruments, this could lead to decreased performance through contamination of the analyzer, although the "dirty" source is rapidly cleaned by simply washing with a cotton swab dabbed in water and then methanol or acetone. Baking the source overnight usually eliminates any of the sputtered liquids. In addition, it is recommended that the source rough pump oil be changed on a regular preventative maintenance schedule. For an instrument which utilizes CF-FAB daily, an oil change may be necessary as often as every month or two, depending on operating conditions. In case of frequent use, daily ballasting of the rough pumps is recommended.

1.5 CONCLUSION

CF-FAB is a valuable tool for the analysis of polar compounds and is well-suited for those dissolved in aqueous solutions. When used in the flow-injection mode of operation, it provides an easy and fast method for the direct injection of sample solutions and lends itself to automated operation. Performance advantages, relative to standard FAB, include increased sensitivity, decreased background, and decreased ion suppression effects. In addition, it can be used as a liquid introduction interface for separation techniques such as LC and CZE, as discussed in detail in other chapters of this book.

REFERENCES

1. R.J. Day, S.E. Unger, and R.G. Cooks, *Anal. Chem. 52*, 557A (1980).
2. D.F. Torgerson, R.P. Skowronski, and R.D. Macfarlane, *Biochem. Biophys. Res. Commun. 60*, 616 (1974).
3. M. Barber, R.S. Bordoli, R.D. Sedwick, and A.N. Tyler, *J. Chem. Soc. Chem. Commun.* 325 (1981).
4. P.A. Lyon, (Ed.) *Desorption Ionization Mass Spectrometry*, Amer. Chem. Soc., Washington, D.C. (1985).
5. M. Barber, R.S. Bordoli, G. Elliott, R.D. Sedgwick, and A.N. Tyler, *Anal. Chem. 54*, 645A (1982).
6. W.D. Lehmann, M. Kessler, and W.A. Koenig, *Biomed. Mass Spectrom. 11*, 217 (1984).
7. J. Sunner, A. Morales, and P. Kebarle, *Intl. J. Mass Spectrom. and Ion Phys. 86*, 169 (1988).
8. S. Naylor, F. Findeis, B.W. Gibson, and D.H. Williams, *J. Am. Chem. Soc. 108*, 6359 (1986).
9. R.M. Caprioli, W.T. Moore, and T. Fan, *Rapid Commun. Mass Spectrom. 1*, 15 (1987).
10. R.M. Caprioli, T. Fan, and J.S. Cottrell, *Anal. Chem. 58*, 2949 (1986).
11. R.M. Caprioli, and T. Fan, *Biochem. Biophys. Res. Commun. 141*, 1058 (1986).
12. R.M. Caprioli, *Anal. Chem. 62*, (1990).
13. Y.Ito, T. Takeuchi, D. Ishi and M. Goto, *J. Chromatogr. 346*, 161 (1985).

14. S.N. Lin, S. Chang, and R.M. Caprioli, *Proceedings of the 38th ASMS Conference on Mass Spectrometry and Allied Topics*, Tucson, AZ, 3-8 June, 1990.
15. R.M. Caprioli, W.T. Moore, G. Petrie, and K. Wilson, *Int. J. Mass Spectrom. Ion Proc. 86*, 187 (1988).
16. M.A. Moseley, L.J. Deterding, J.S.M. deWit, K.B. Tomer, R.T. Kennedy, N. Bragg, and J.W. Jorgenson, *Anal. Chem. 61*, 1577 (1989).
17. S.J. Gaskell, and R. S. Orkiszewski, Chapter 2, this volume.
18. H.B. Bull and K. Breese, *Arch. Biochem. Biophys. 161*, 665 (1974).

Chapter 2

TRACE ANALYSIS

Simon J. Gaskell and Ralph S. Orkiszewski

The introduction of fast atom bombardment and liquid secondary ion mass spectrometry (1) significantly broadened the scope of application of mass spectrometry to the biomedical sciences. Analytes intractable to other methods of ionization (such as electron ionization or chemical ionization), because of their polarity or relatively high molecular mass, were amenable to ionization using FAB. The extensive literature which has developed on FAB MS has been largely focused on applications in which the emphasis is on structural characterization, rather than on the detection and quantitation of trace components of mixtures. The few exceptions to this general observation have included work from our laboratory on the determination of a steroid sulfate (2) and of a platelet-activating factor (3), a phospholipid mediator, using FAB and tandem mass spectrometry. In both cases, selective detection of analyte and internal standard used limited mass range parent ion scanning.

The recent innovation of CF-FAB for continuous sample introduction for FAB MS (4,5) has several potential advantages in the area of trace analysis:

(a) Temporal definition of the signal attributable to the analyte. Conventional sample introduction for FAB MS does not permit the distinction of analyte and background signal during a single analysis. Thus, for example, during a quantitative analysis of a cortisol derivative by FAB/tandem MS using dual reaction monitoring (6), no estimate could be made of any background contribution to the detected signals. In contrast, CF-FAB analysis can achieve a stable baseline signal prior to, and following, the emergence of the signal associated with the analyte.

(b) Reduction of suppression effects. The application of FAB MS to mixture analysis has been hampered by the disproportionately sensitive detection of components with high surface activity (7,8). Thus, it has been observed that minor components of mixtures may be obscured by the presence of other constituents of higher surface activity (8,9). The continuous and vigorous refreshment of the surface achieved during analyses with CF-FAB might be expected to reduce this effect, and indeed it has been observed that suppression effects and variations in relative sensitivities are less pronounced using continuous flow, rather than conventional, sample introduction (10).

(c) Generation of a short-lived, intense signal. The longevity of the signal obtained during FAB MS with conventional sample introduction has been considered a virtue in promoting the feasibility of mass analyzed ion kinetic energy and other variants of the tandem MS experiment (11). Again, the emphasis has been on utility for structure elucidation purposes. When trace analysis is the objective, the advantages of a short-lived signal of maximum intensity are clear. Early reports of the use of CF-FAB suggested that increased absolute signal intensities were observed (12), although the apparent sensitivity advantage of the technique is derived, at least in part, from the reduction in background signal (see below).

(d) Reduced background signal associated with the liquid matrix. The practical simplicity of the std-FAB technique is largely associated with the use of a liquid matrix, whose principal function is to facilitate the replenishment of the concentration of analyte at the surface exposed to the primary atom beam. In the continuous flow technique, renewed exposure of analyte to the primary beam is achieved by the flow itself and, in practice, a proportion of matrix compound of only a few per cent is required for satisfactory operation. (This is discussed in detail below.) The background signal is accordingly decreased and the ratio of analyte response to background is improved (10).

(e) Feasibility of direct coupling of liquid chromatography with FAB MS. Clearly, the facility for mixture and trace analysis is significantly improved if an on-line separation technique is incorporated. The potential benefits of liquid chromatography LC/CF-FAB MS in trace analysis parallel the advantages of, for example, gas chromatography-electron impact MS and LC thermospray MS, but the range of applicability of coupled chromatography MS is substantially increased. The earliest report of continuous fluid introduction to a FAB ion source (4) included on-line high-performance liquid chromatographic separations. A large number of reports of LC/CF-FAB MS have since appeared, together with a few descriptions of combined capillary zone electrophoresis CF-FAB MS (13,14).

The emphasis of our laboratory on the detection and quantitation of trace components of complex mixtures of biological origin has encouraged us to explore the use of CF-FAB in this area. In this chapter, we describe preliminary studies to optimize detection during CF-FAB with respect to both sensitivity and selectivity of detection. One of the major advantages of mass spectrometry as a means of detection of trace components is the selectivity achievable, with benefits both in the confidence of identification and the accuracy of quantitation of trace components. When tandem MS is employed, further increases in selectivity of detection may be achieved; we describe such applications here.

2.1 INSTRUMENTATION AND METHODS

Analyses were performed using a VG ZAB SEQ instrument which has the configuration, BEqQ (where B = magnetic sector, E = electric sector, q = rf-only quadrupole collision cell, and Q = quadrupole mass filter). The instrument was equipped with a standard VG FAB ion source designed for conventional or for continuous flow introduction of sample. The source heater was adjusted to achieve a stable signal for a particular solvent composition during continuous flow operation. The primary beam was neutral xenon with an energy of approximately 8 keV.

Three modes of data acquisition were employed, each under the control of the VG 11-250 data system. The majority of the analyses used selected ion monitoring with detection following the electric sector. The mass spectrometric resolution (10% valley definition) was 800 or 1500. For selected reaction monitoring (used in the analysis of a platelet-activating factor), precursor ions were selected (at a resolution of 800) using the double-focusing portion of the instrument and allowed to fragment in q (without added collision gas). The quadrupole mass filter, Q, was set to transmit the appropriate product ion (resolution 1-1.5 mass units). Dual reaction monitoring involved synchronized switching of accelerating and quadrupole voltages. Product ion scanning (during analyses of leucine-enkephalin) involved selection of $(M+H)^+$ ions using the double-focusing portion of the instrument (resolution 800), collisionally activated decomposition (CAD) in q, and scanning of Q (set to a resolution of 2-3 mass units). CAD used argon as collision gas, at an estimated pressure in q of 2×10^{-4} mbar. The collision energy (laboratory frame-of-reference) was 26 eV. Acquisition was in the "multichannel analyzer" mode. The scan rate was 10 s/scan and 15 scans were accumulated preceding and during the elution of leucine-enkephalin.

Solvent was delivered to the CF-FAB probe at a rate of 4.5 μl/min using a pair of Waters prototype microstep-driven syringe pumps, each capable of delivering up to 30 μl/min under the control of a modified signal generated by a Waters Model 680 Automated Gradient Controller. Sample injections were made via a Rheodyne Model 7520 injector fitted with a 0.5 μl fixed volume internal injection loop. A fused silica capillary (50 μm i.d.) was connected directly to the outlet port of the injector, using a stainless steel ferrule and a small piece of Teflon tubing with an internal diameter close to the outer diameter of the fused silica.

The CF-FAB probe was of standard VG Analytical design (purchased in 1988), slightly (but critically) modified by the elimination of space between the 50 μm fused silica sample delivery capillary and the internal surface of the stainless steel probe tip (Figure 2.1). This was achieved by wrapping a small

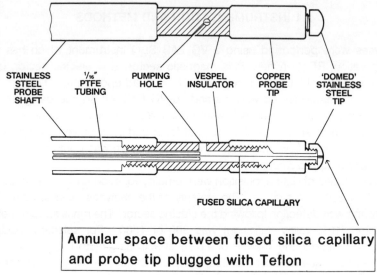

Figure 2.1. Design of the VG Analytical CF-FAB probe and the modification incorporated for this work.

strip of Teflon tape around the capillary and screwing the probe tip firmly into position. The excess Teflon was shaved away and the capillary was recessed slightly from the probe tip. Both flat and hemispherical probe tips have been employed with no significant differences observed with regard to peak shape or sensitivity.

Figure 2.2 shows schematically the configuration for CF-FAB established for this work. Three modes of operation are possible, involving direct introduction of sample solution via suction from a reservoir, loop injection into a controlled flow of solvent, and on-line LC/MS. The examples reported here concern only the loop injection method, which was used to optimize the technique for subsequent on-line chromatographic separations.

Several features of the plumbing were critical to the achievement of stable analyte responses with minimal band broadening. Superior stability was achieved using 50 μm, rather than 75 μm, internal diameter fused silica capillary between the injector and the probe tip. Band broadening was

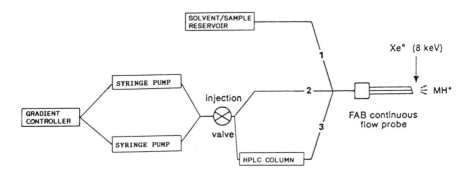

Figure 2.2. Instrument operational configurations for CF-FAB MS analyses. Details of the components are given in the text.

critically affected by the coupling of the fused silica tube to the injector port; the union incorporating a ferrule/Teflon combination (described above) was found to be effective. Stability and sample peak width were both substantially improved by filling of the void between the fused silica capillary and the internal surface of the sample probe tip; indeed, satisfactory operation was not achieved without this modification. With the experimental arrangement adopted, injection of sample volumes of 0.5 μl into a solvent flow of 4.5 μl/min gave peaks with a typical width of 10-15 seconds at 10% of peak height.

2.2 OPTIMIZATION OF EXPERIMENTAL PARAMETERS

Figure 2.3 shows repeated analyses of the peptide leucine-enkephalin (10 ng/0.5 μl water) with selected ion monitoring (SIM) of $(M+H)^+$ ions of m/z 556. The solvent incorporated 10% glycerol and 1% saturated aqueous oxalic acid. Duplicate injections were performed at each of the four indicated source temperatures. Optimal results were obtained at 70 °C. The data indicate the effect of source temperature on response but the apparent optimal temperature should not be extrapolated to other source or inlet designs. The expedient adopted in our work of Teflon plugging of the void between the fused silica capillary and the probe tip presumably introduces a significant thermal gradient in this region so that the temperature at the exit of the fused silica capillary is unknown.

The incorporation of 1% saturated aqueous oxalic acid in the eluting solvent was found to give maximum responses during SIM of leucine-enkephalin and

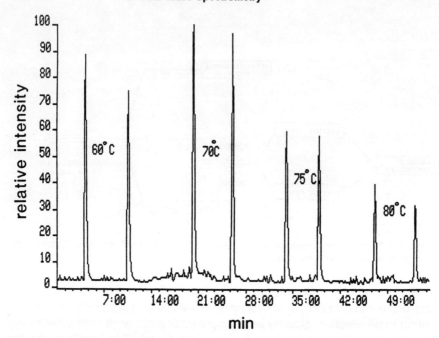

Figure 2.3. Replicate injections of leucine-enkephalin (10 ng in 0.5 μl water) with SIM of m/z 556.28. Two injections were made at each of four ion source temperatures, as indicated. The solvent was acetonitrile/water/glycerol/saturated aqueous oxalic acid (45/44/10/1) at a flow rate of 4.5 μl/min. The MS resolution was 800.

other peptides; increased proportions gave no significant increase in response.

Using an eluting solvent of acetonitrile/water/glycerol/saturated aqueous oxalic acid (45/44/10/1), intense and reproducible responses to injections of 500 pg leucine-enkephalin were obtained (Figure 2.4) but significant responses were also observed for injection of water alone. It was established that the blank response was associated with a general elevation of background signal, rather than carry-over of analyte from previous injections. Figure 2.5(a) indicates that the signal attributable to glycerol background at m/z 556 is significant. The figure represents the set-up page used for source tuning via the VG 11-250 data system and indicates that, in addition to the major glycerol background peak of m/z 553 ([glycerol]$_6$+H$^+$), a minor peak is observed with an exact mass close to that of the (M+H)$^+$ ion of leucine-enkephalin (m/z 556.28). Figure 2.5(b) shows equivalent data obtained during CF-FAB of leucine-enkephalin when the eluting solvent incorporated 1% 2,2'-dithiodiethanol as the matrix additive. The mass deficiency of 2,2'-dithiodiethanol-derived ions shifts background peaks so that a mass spectrometric resolution of 1500 is adequate to separate the (M+H)$^+$ ions of leucine-enkephalin from

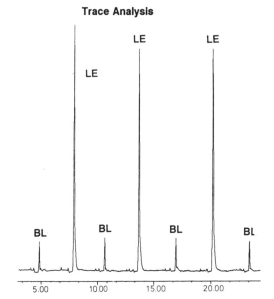

Figure 2.4. Replicate injections of leucine-enkephalin (LE) (500 pg in 0.5 μl water) with alternate "blank" injections of water alone (BL). Ions of m/z 556.28 were monitored. The solvent was acetonitrile/water/glycerol/saturated aqueous oxalic acid (45/44/10/1) at a flow rate of 4.5 μl/min. The MS resolution was 800.

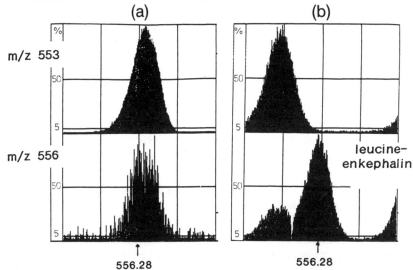

Figure 2.5. Peak profiles of ions of m/z 553 and 556 during analysis of (a) a solution of glycerol and (b) a sample of leu enkephalin by CF-FAB using acetonitrile/water (1/1) containing 1% saturated aqueous oxalic acid and 0.5% 2,2'-dithiodiethanol at a flow rate of 4.5 μl/min. The MS resolution was 1500.

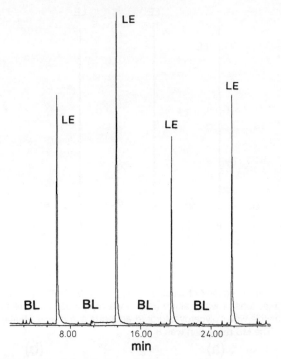

Figure 2.6. Replicate injections of leucine-enkephalin (LE) (500 pg in 0.5 µl water) with alternate "blank" injections of water alone (BL). Ions of m/z 556.28 were monitored. The solvent was acetonitrile/water (1/1) with 1% saturated aqueous oxalic acid and 0.5% 2,2'-dithiodiethanol at a flow rate of 4.5 µl/min. The MS resolution was 1500.

the background as shown in Figure 2.6. This has the effect of eliminating the general background rise recorded when sample is bombarded on the target, a procedure of critical importance when SIM recordings are being made. This is clearly seen by comparing the spectra shown in Figures 2.4 and 2.6 obtained from the analysis of 500 pg of leucine-enkephalin. Blank injections made at an instrument resolution of 800 (Figure 2.4) showed a significant peak at m/z 556 which was derived from the matrix material, while that obtained at a resolution of 1500 (Figure 2.6) shows this background almost completely eliminated.

The likelihood of background interference also can be further diminished by the reduction in the proportion of "matrix" in the eluting solvent. For the SIM analysis of leucine-enkephalin (500 pg) shown in Figure 2.6, the solvent system was composed of acetonitrile/water (1/1) containing 1% saturated aqueous oxalic acid and 0.5% 2,2'-dithiodiethanol. The combined effect of higher instrument resolution and low concentration of matrix to lower the background is not unique to this sample and can be applied to the majority of compounds of biological interest.

While the choice of matrix additive had a demonstrable effect on baseline stability and helped eliminate "blank" responses, absolute analyte responses were little affected by the matrix selected. This suggests that the principal function of the matrix is to promote signal stability by controlled solvent volatilization at the sample probe tip rather than to replenish surface concentrations of analyte.

In summary, the optimization experiments suggested the following standard conditions for trace analyses:

source temperature: 70 °C
solvent composition: acetonitrile/water (1/1), containing 1% of saturated
  aqueous oxalic acid and 1% 2,2'-dithiodiethanol
flow rate: 4.5 μl/min
mass spectrometric resolution: 1500 (10% valley)
scanning mode: SIM

Under these experimental conditions, SIM analyses of leucine-enkephalin $(M+H)^+$ ions gave a low picogram detection limit (Figure 2.7). Excellent

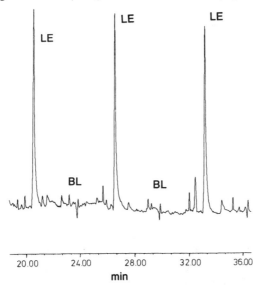

Figure 2.7. Replicate injections of leucine-enkephalin (LE) (25 pg in 0.5 μl water) with alternate "blank" injections of water alone (BL). Ions of m/z 556.28 were monitored. The solvent was acetonitrile/water (1/1) with 1% saturated aqueous oxalic acid and 1.0% 2,2'-dithiodiethanol at a flow rate of 4.5 μl/min. The MS resolution was 1500.

sensitivities of detection were also observed for larger peptides, such as bradykinin (M_r 1059) as shown in Figure 2.8 for the detection of 500 pg with SIM of $(M+H)^+$ ions.

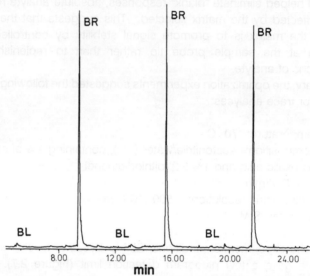

Figure 2.8. Replicate injections of bradykinin (BR) (500 pg in 0.5 µl water) with alternate "blank" injections of water alone (BL). Ions of m/z 1060.57 were monitored. The solvent was acetonitrile/water (1/1) with 1% saturated aqueous oxalic acid and 1.0% 2,2'-dithiodiethanol at a flow rate of 4.5 µl/min. The MS resolution was 1500.

2.3 CF-FAB/TANDEM MS OF LEUCINE-ENKEPHALIN

Tandem MS permits improved structural characterization and superior selectivity of detection of trace components of complex mixtures. For example, we have used std-FAB (with conventional sample introduction) and hybrid tandem MS to confirm the presence of leucine-enkephalin in an extract of canine brain (15). However, product ion spectra obtained for trace components may be obscured by fragments derived from isobaric precursor ions associated with the matrix. We have therefore determined the analytical limit for recording a low-energy collisionally activated decomposition product ion spectrum of leucine-enkephalin $(M+H)^+$ ions during CF-FAB analysis. Figure 2.9 shows the spectrum obtained by analysis of 1 ng (ca 2 pmol) of leucine-enkephalin. Prominent ions are attributable to the fragmentations shown in Figure 2.10. An additional abundant product ion of m/z 402 derives from the matrix background by loss of a monomeric unit of 2,2'-dithiodiethanol.

Figure 2.9. Product ion (MS/MS) spectrum obtained by low-energy CAD of $(M+H)^+$ ions derived from leucine-enkephalin. The spectrum was obtained using 1 ng (*ca.* 2 pmol) of analyte introduced by CF-FAB. Analytical conditions were as described in Figure 2.8.

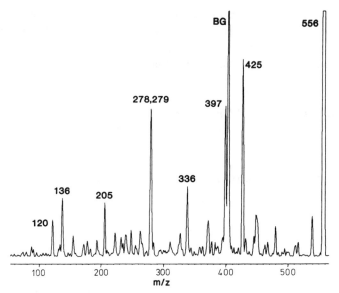

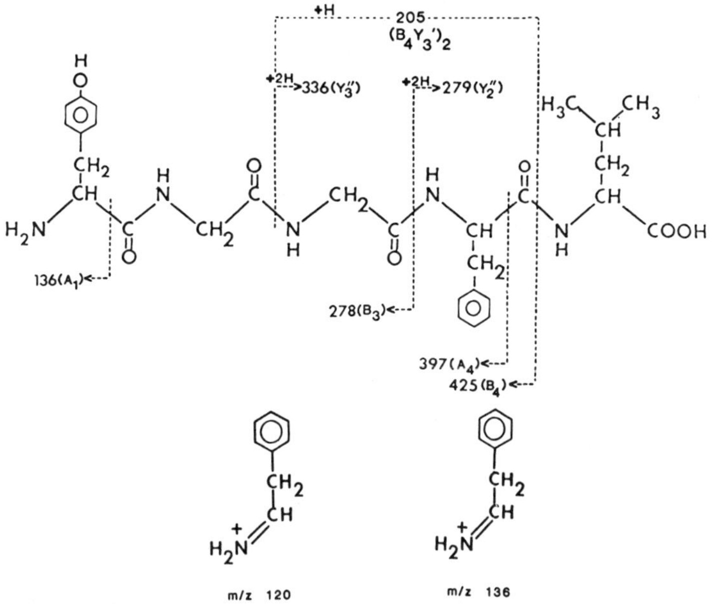

Figure 2.10. Fragmentations of leucine-enkephalin $(M+H)^+$ ions.

This reflects in part the accumulation of product ion scans preceding the emergence of leucine-enkephalin from the transfer capillary. For this analysis, the precursor ion resolution was 800 and, therefore, the background ion at m/z 556 was also collisionally decomposed and the product ions recorded.

2.4 DETERMINATION OF PLATELET-ACTIVATING FACTOR (PAF) IN CELL EXTRACT

PAF (1-O-alkyl-2-acetyl-sn-glycero-3-phosphocholine) is a unique biologically active phospholipid which is produced by several cell types and which mediates a host of inflammatory reactions and pathophysiological processes (16). Analyses for PAF, using routine methods such as bioassay or radioimmunoassay, are complicated by the heterogeneity observed in the alkyl substituent. Gas chromatography-mass spectrometry has been applied to the determination of individual molecular species following cleavage of the phosphocholine head group and derivatization (17,18). We recently described (3) FAB/tandem MS analyses of intact PAF in which quantitation was achieved by limited mass range parent ion scanning to detect the principal fragmentation of hexadecyl-PAF and of the [$^2H_3$]analogue used as an internal standard (Figure 2.11). In the present work, we have exploited the advantages

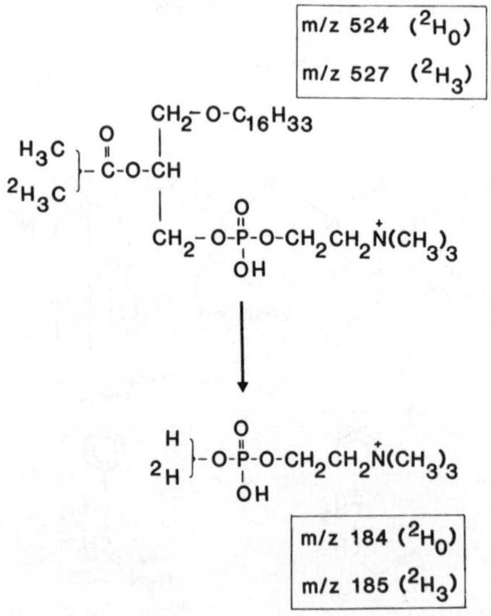

Figure 2.11. Structure and principal fragmentation of hexadecyl-PAF and the [$^2H_3$]-labelled analogue.

of continuous flow sample introduction to determine hexadecyl-PAF using selected reaction monitoring analyses of cell extracts with MS/MS.

Stimulation and extraction of human neutrophils was performed using a variation of the method previously described (3). Briefly, neutrophils were stimulated by addition of calcium ionophore and extracted by the method of Bligh and Dyer (19). A phospholipid fraction was obtained by separation using an aminopropyl-silica cartridge and further separated by normal phase HPLC, according to Blank and Snyder (20). [^{2}H$_3$]acetyl PAF was used as the internal standard.

Figure 2.12 shows such analyses of extracts of stimulated human neutrophils following HPLC fractionation. The accelerating voltage and

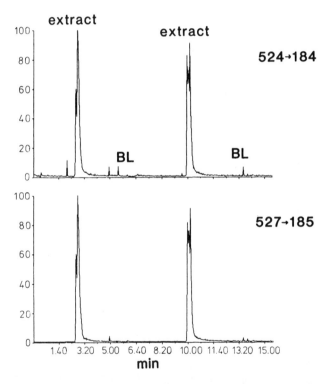

Figure 2.12. CF-FAB analyses of extracts of stimulated neutrophils with selected reaction monitoring of (m/z 524 → 184) and (m/z 527 → 185), characteristic of hexadecyl-PAF and the [^{2}H$_3$]-analogue, respectively. The solvent was acetonitrile/water (1/1) with 1% saturated aqueous oxalic acid and 1.0% 2,2'-dithiodiethanol at a flow rate of 4.5 µl/min. The MS resolution was 1500. Each selected reaction monitoring trace is independently normalized in the figure; however, the response ratio for hexadecyl-PAF/[^{2}H$_3$]hexadecyl-PAF was approximately 0.6.

quadrupole mass filter voltages were switched in synchronization to permit the alternate detection of the m/z 524→ 184 transition, characteristic of hexadecyl-PAF, and the m/z 527→ 185 transition characteristic of the [^{2}H$_3$]hexadecyl-PAF internal standard. The selected reaction monitoring traces shown in Figure 2.12 are individually normalized; separate determination of peak heights indicated a response ratio, hexadecyl-PAF/internal standard, of approximately 0.6. By reference to the analysis of standard mixtures, this indicated a total of about 60 ng hexadecyl-PAF, corresponding to a concentration of approximately 2.6 ng/10^6 cells. Dual selected ion monitoring of m/z 524 and 527 during analyses of the same extracts by CF-FAB with single MS detection gave a significantly higher peak height ratio of 0.8. This result is consistent with our earlier finding that conventional (single) MS detection is of inadequate selectivity for the accurate determination of PAF in cell extracts.

2.5 CONCLUSION

The data reported here indicate that careful optimization of experimental parameters allows CF-FAB MS to be used effectively for trace analysis. Stable signals may be achieved with a 1% or lower proportion of matrix added to the solvent. The use of 2,2'-dithiodiethanol as a matrix additive, in conjunction with slightly elevated mass spectrometric resolution (1500), is beneficial in separating background from analyte signals. As observed with trace analyses using other mass spectrometric methods, the incorporation of tandem MS techniques with CF-FAB provides improved selectivity of detection.

ACKNOWLEDGEMENTS

This work was supported by the National Institutes of Health (GM 34120 and AI 26916) and by a generous gift from the Burroughs Wellcome Company. The VG ZAB SEQ mass spectrometer was purchased by the Howard Hughes Medical Institute. The authors thank Serenella Rotondo for the preparation of the extract of stimulated neutrophils, and Gareth Thorne and Theodore Dourdeville for their help in obtaining some of the data presented in this chapter.

REFERENCES

1. M. Barber, R.S. Bordoli, G.J. Elliott, R.D. Sedgwick and A.N. Tyler, *Anal. Chem. 54*, 645A (1982).
2. S.J. Gaskell, *Biomed. Environ. Mass Spectrom. 15*, 99 (1988).
3. P.E. Haroldsen and S.J. Gaskell, *Biomed. Environ. Mass Spectrom. 18*, 439 (1989).
4. Y. Ito, T. Takeuchi, D. Ishi and M. Goto, *J. Chromatogr. 346*, 161 (1985).
5. R.M. Caprioli, T. Fan and J.S. Cottrell, *Anal. Chem. 58*, 2949 (1986).
6. S.J. Gaskell, in *Clinical Chemistry, An Overview*, edited by N.C. den Boer, C. van der Heiden and J.H.M. Souverijn, p. 589, Plenum Press, New York, 1989.
7. W.V. Ligon and S.B. Dorn, *Int. J. Mass Spectrom Ion Proc. 57*, 75 (1984).
8. S. Naylor, A.F. Findeis, B.W. Gibson and D.H. Williams, *J. Amer. Chem. Soc. 108*, 6359 (1986).
9. M.R. Clench, G.V. Garner, D.B. Gordon and M. Barber, *Biomed. Mass Spectrom. 12*, 355 (1985).
10. R.M. Caprioli, W.T. Moore and T. Fan, *Rapid Commun. Mass Spectrom. 1*, 15 (1987).
11. M. Barber, R.S. Bordoli, R.D. Sedgwick and A.N. Tyler, *Biomed. Mass Spectrom. 9*, 208 (1982).
12. R.M. Caprioli and T. Fan, *Biochem. Biophys. Res. Commun. 141*, 1058 (1986).
13. J.S. de Wit, L.J. Deterding, M.A. Moseley, K.B. Tomer and J.W. Jorgensen, *Rapid Commun. Mass Spectrom. 2*, 100 (1988).
14. Chapters 5 and 6, this volume.
15. S.J. Gaskell, M.H. Reilly and C.J. Porter, *Rapid Commun. Mass Spectrom. 2*, 142 (1988).
16. P. Braquet, L. Touqui, T.Y. Shen and B.B. Vargaftig, *Pharmacol. Rev. 39*, 97 (1987).
17. K. Satouchi, M. Oda, K. Yasunage and K. Saito, *J. Biochem. 94*, 2067 (1983).
18. C.S. Ramesha and W.C. Pickett, *Biomed. Mass Spectrom. 13*, 107 (1986).
19. E.G. Bligh and W.J. Dyer, *Can. J. Biochem. Physiol. 37*, 911 (1959).
20. M.L. Blank and F.Snyder, *J. Chromatogr. Biomed. Applic. 273*, 415 (1983).

Chapter 3

QUANTITATIVE ANALYSIS

S. P. Markey and Ming-Chuen Shih

The quantitation of polar compounds in biological fluids such as sulfate or glucuronide conjugates of drugs and trace quantities of endogenous metabolites is a major interest in our laboratory, and we have sought to utilize continuous-flow (CF)-liquid secondary ion mass spectrometry (LSIMS) for this purpose. Principally, we wanted simpler means of chemical analysis in order to obtain increased accuracy for quantification of intact compounds, avoiding the inherent problems associated with converting polar compounds into gas phase stable derivatives. The advantages of CF are very significant for quantitative FAB or LSIMS. Sample introduction into the ion source in a defined time window, as compared to the static probe condition, permits mass spectrometric data analysis using the standard chromatographic procedures such as background subtraction and peak area integration. These are routinely employed for quantitative selected ion monitoring by gas chromatography-mass spectrometry, where the use of stable isotopomer internal standards has made quantitative analyses facile. The choice of LSIMS versus FAB was made on the basis of emerging data regarding cesium gun equivalence with fast atom sources, with the added observation that cesium guns do not impose the gas pressure burden which saddle field fast atom sources add to the instrument's pumping load. Thus, the data and operational observations described in this chapter were derived from CF-LSIMS rather than CF-FAB ionization techniques, although there is an equivalence of data obtained by either method. Consequently, experimental conditions reported for CF-LSIMS are readily transferable to CF-FAB and vice versa. This chapter summarizes procedures and data which we have previously reported (1-3), as well as studies in progress.

3.1 INSTRUMENTATION

The instrumentation used in our laboratory for the CF-LSIMS experiments has been constantly evolving, and this chapter summarizes the rationale for the changes as enumerated in the following text. All of the experiments have been conducted using a Finnigan MAT TSQ 70 triple-stage quadrupole mass spectrometer equipped with a Finnigan MAT prototype BioProbe™ ion source. Liquid is delivered to the stationary probe target as shown schematically in Figure 3.1. Unlike most of the commercial probe designs in which a fused

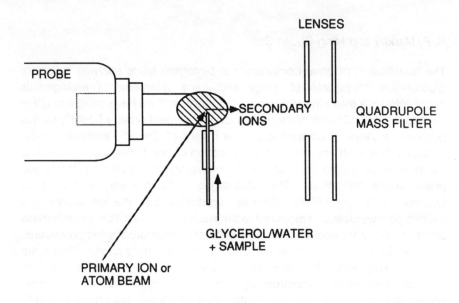

Figure 3.1. Schematic representation of Finnigan BioProbe CF-LSIMS arrangement for the TSQ 70. A conventional direct insertion probe is utilized with the liquid matrix delivered through an alternative fused silica capillary tube inlet to the probe target surface.

silica capillary in the probe shaft delivers liquid through the probe tip, this design utilizes a conventional probe and delivers the liquid through an alternative capillary inlet to the probe target surface. This is perhaps a less expensive manufacturing alternative to the probe shaft design, but which initially appeared to satisfy all of the requirements of CF-FAB. However, the disadvantage of this design is the precise mechanical positioning of the probe shaft with respect to the contacting capillary tube which is required in order to obtain the desired liquid film condition.

Glycerol/water solutions (20/80, v/v, unless otherwise noted) were pumped at flow rates of about 6-8 μl/min by a Gilson Model 302 HPLC pump with a microflow adaptor and pulse dampener (Model 802B). In order to provide sufficient resistance for the pump to regulate, a microbore HPLC column was placed prior to the sample injector. Samples were introduced with a 0.5-µl loop injector (Rheodyne, Model 7410). Transfer line tubing was 75 µm (i.d.) fused silica (Polymicro Tech.). The thermal energy required to keep the liquid film stable and avoid freezing from evaporative heat loss was supplied either by direct contact of the fused silica capillary with an aluminum guide tube attached to the heated source block, and the heat content of the probe. Temperature measurements of the ion source block, while not an accurate index of the liquid film temperature, were generally at 40-45 °C.

The TSQ 70 instrument vacuum housing is a two-compartment aluminum rectangular box with O-rings providing the vacuum seals to a large glass cover. Because we anticipated the need to have frequent access to the ion source compartment of the instrument, but not the quadrupole rod and analyzer compartment, the single glass lid was cut into two parts. Thus the smaller glass cover over the source compartment is readily removed. An external viewing telescope was positioned over the probe tip area to provide expanded detail of the probe tip film. The position of the probe tip is critical, both for film stability and for optimum secondary ion emission as shown in Figure 3.2.

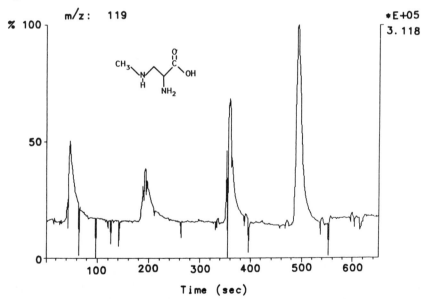

Figure 3.2. The effect of probe alignment relative to the fused silica capillary is critical, as indicated by monitoring the secondary ion current from repetitive injections of the same quantity of an amino acid sample with minor adjustment to probe position between each injection.

An unfocused cesium ion gun (Phrasor Scientific) was used in the first series of experiments described here. It was operated at 5-6 keV accelerating potential, and a heater current setting to deliver approximately 1-5 $\mu A/cm^2$, as estimated from measurements described later in this chapter. This gun was modified by adding a cylindrical focusing lens system, and was then replaced with a prototype Phrasor focused cesium gun.

3.2 PARAMETERS AFFECTING QUANTITATION

Several series of tests have been performed to evaluate the performance of this instrumentation and its applicability for routine assay. We have used compounds being assayed in our laboratory by gas phase techniques, so that we have suitable isotopomer internal standards, and a basis of comparison with another method for sensitivity and reproducibility. Two such compounds are 1-methyl-4-phenylpyridinium (MPP^+), and L-tryptophan. A series of

MPP+   L-TRYPTOPHAN

repetitive loop injections of solutions of L-tryptophan of constant concentration containing a varying concentration of $^2H_5$-L-tryptophan produced the selected ion signals shown in Figure 3.3. When the ratio of signals from the respective $(M+H)^+$ ions was plotted versus the concentration of $^2H_5$-L-tryptophan, a linear standard curve resulted, suitable for quantitative analyses. The reproducibility of repetitive measurements is excellent. For example, when a mixture containing 50 ng MPP^+ and 125 ng $^2H_3$-MPP^+ per 0.5 μl was repetitively injected, a mean ratio of 0.3983 (calc. 0.40)±.002 s.d. (range 0.3955-0.4011, $n=10$) was determined for symmetrical signals (50 seconds wide at base line, 15 seconds wide at half height) at m/z 170 and 173. This experiment tested the precision but not the accuracy of the determination of the expected weight ratio. As in the tryptophan example, quantitative analyses of MPP^+ using $^2H_3$-MPP^+ as internal standard was demonstrated with a linear standard curve with a correlation coefficient of 0.999 over the range 200 pg to 200 ng. A

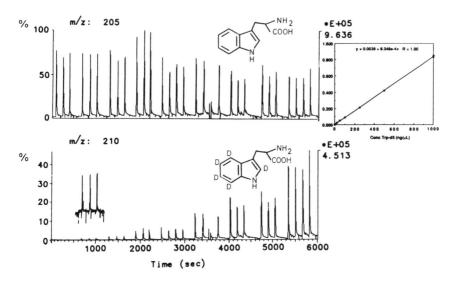

Figure 3.3. Repetitive injections of solutions containing a constant quantity of L-tryptophan and increasing quantities of $^2H_5$-tryptophan. A plot of the ratio of non-deuterated to deuterated isotopomer signals versus concentration of deuterated tryptophan produced a linear standard curve over the range 5 to 1000 ng. The 5 ng response has been scale expanded for the m/z 210 plot in order to indicate the signal/noise ratio.

detection limit of 200 pg was routinely possible with a signal-to-noise ratio of 3:1. As in other chromatographic techniques, obtaining linearity at maximum sensitivity and detectability requires recognition of the surface properties of the analyte and the materials through which it must pass. In the case of MPP$^+$, deactivation of the surfaces of the transfer lines and injection valve could be accomplished by several means; the inclusion of a structural analogue, homologue, or isotopomer in the mobile phase, or saturation of the active surface sites by injection of a large bolus of analyte prior to quantitative runs.

Considering the difficulty in separating analyte from matrix signals at low m/z values in conventional static FAB, the results obtained with relatively pure standards illustrate the promise of CF-LSIMS. However, these examples of successful analyses have not yet been translated into routine analyses of the same or similar compounds when extracted or concentrated from biological sources. Several problems have been encountered with the application of CF-LSIMS to such samples, although the ease of quantitative analysis of relatively clean samples is notable.

One problem is that of non-specific background, as observed in the semiquantitative analysis of MPP$^+$ from adrenal tissue (Figure 3.4). In this

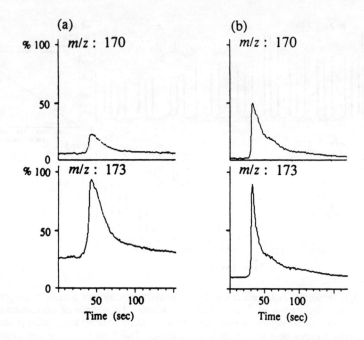

Figure 3.4. Reconstructed ion recordings at m/z 170 (MPP$^+$) and 173 (^{2}H$_3$-MPP$^+$) for adrenal gland extracts prepared (a) from control rats and (b) from rats sacrificed 4 hours after injection of MPTP, 10 mg/kg (subcutaneous). A fixed amount of ^{2}H$_3$-MPP$^+$ was added as internal standard, and signals at m/z 170 normalized to that at m/z 173 for comparison.

example, the same quantity of ^{2}H$_3$-MPP$^+$ was added to control (a) and drug-treated (b) extracts. Thus, any signal measured at M$^+$ for MPP$^+$ (m/z 170) in Figure 3.4(a) is non-specific background due to other tissue components, whereas the increased response in (b) at m/z 170 is due to the presence of MPP$^+$. As is seen, there is a real increase in chemical noise due to the complex and crude biological extract which is not characteristic of pure MPP$^+$ isotopic standards.

Several solutions to this problem are possible, one of which is the use of chemical separation techniques. Another is the use of the increased specificity obtained by tandem MS (MS/MS) with CF sample introduction. As in the above example cited, CF-LSIMS was employed for the quantitation of 2-amino-3-(methylamino)propanoic acid (BMAA) in plant extracts. The direct analysis of the (M+H)$^+$ from BMAA and its ^{2}H$_3$-isotopomer exhibited the same non-specific background signal increase as observed in the analysis of MPP$^+$.

In fact, most procedural blanks from any sample work-up procedure exhibit general "peak-at-every-mass" background. A significant sample clean-up and enrichment could be realized by formation of the ethyl ester of BMAA and its extraction into organic solvents, but the non-specific background signal precluded quantitative measurements. BMAA-ethyl ester, $(M+H)^+$ 147, exhibited a useful collisionally activated decomposition (CAD) fragment at m/z 44 which was shifted to m/z 47 in the $^2H_3$-isotopomer internal standard (Scheme 3.1). By alternating between m/z 147-to-44 and 150-to-47

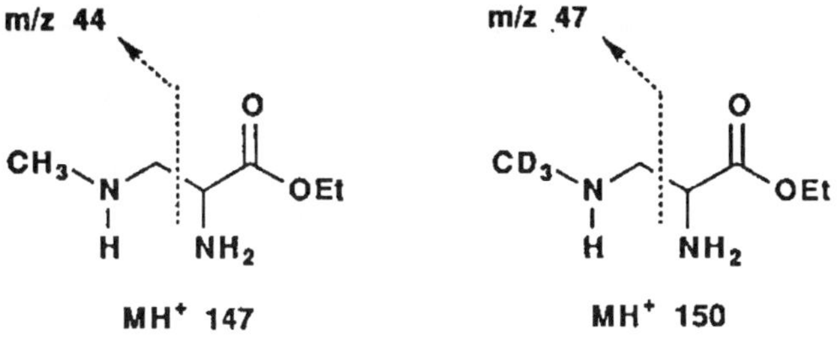

Scheme 3.1. Structures of BMAA-Ethyl ester isotopomers and the origin of the major fragments when the protonated molecular ions from LSIMS are collisionally dissociated.

parent-to-daughter fragmentations, a significant element of specificity was introduced into the detection. Thus, the standard curve obtained with pure standards was readily applied to plant extracts. Examples of selected reaction monitoring for two plant species are shown in Figure 3.5. Note that the extract of the species *Macrozamia lucida* does not contain BMAA-ethyl ester, and that there is a corresponding absence of signal for m/z 147-to-44, indicating that there is no chemical background noise problem. Further, we have compared CF-LSIMS MS determinations of plant extracts with those obtained by GC/MS of trifluoroacylated-BMAA-ethyl ester, and found excellent agreement. Thus, the added specificity of MS/MS is, as expected, approximately equivalent to that gained by chromatography. With both GC/MS and CF-LSIMS assays for BMAA available, we have found that each approach has its own advantages. CF-LSIMS MS offers a significant reduction in both preparation and run time per sample, but these advantages are offset by lower sensitivity (nanograms for CF-LSIMS versus picograms for GC/MS) and a fully automated GC/MS sample introduction capability.

A second problem with biological extracts is suppression of secondary ions due to the presence of salts or organic compounds other than the analyte of interest. This problem has been recognized by practitioners of static FAB MS, who routinely utilize off-line HPLC (for example, see references 4-6), a short

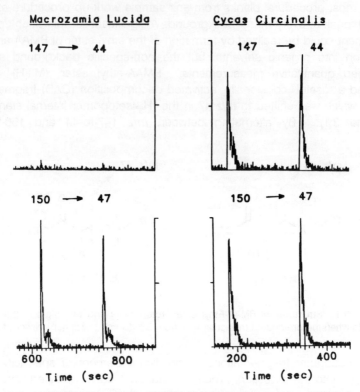

Figure 3.5. CF-LSIMS MS analyses of BMAA-Et indicate the absence of BMAA in extracts obtained from Macrozamia relative to the amount detected in Cycas, both spiked with identical amounts of $^2H_3$-BMAA prior to extraction and derivatization. Note the absence of signal in the 147-to-44 in the top-left chromatogram, indicating the specificity of MS/MS.

desalting column (7), or the addition of surfactants (8) to alleviate this problem. For practitioners of protein sequencing by FAB MS, the separation of peptide digests by HPLC has been critical in that it provides, even under poor resolving conditions, a mixture of peptides with similar hydrophobicities, permitting detection of each molecular species without signal suppression (4-6). Undoubtedly, an optimal approach for CF/LSIMS of peptides is its combination with some on-line separation technique, either microbore HPLC, capillary zone electrophoresis or countercurrent chromatography, in order to ensure detection of all species. To date, we have tried to use off-line clean-up procedures, followed by direct solvent loop injections in order to optimize MS operation time. In several instances, we have been unable to remove contaminating substances from brain or urine extracts. For example, a brain extract which was known (from HPLC-based data) to contain a concentration of MPP^+ which should have been readily detectable by static LSIMS, produced

Quantitative Analysis

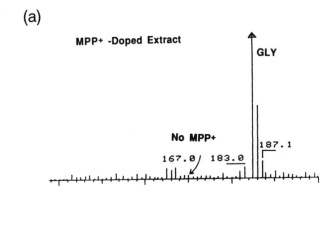

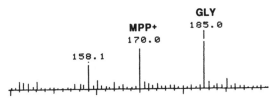

Figure 3.6. Static LSIMS spectra of brain tissue extracts known to contain MPP$^+$. In panel (a) there is very little signal present at m/z 170 relative to the glycerol background. However, in panel (b), after the addition of a trace amount of sodium dodecyl sulfate, the relative intensity of m/z 170 becomes prominent.

a spectrum (Figure 3.6(a)) which was dominated by glycerol ions, and devoid of the expected signal at m/z 170. We have tried the desalting columns recommended by Moon and Kelley (7), but have found our samples to be too polar to separate from the contaminating substances. When sodium dodecyl sulfate (SDS) was added to the glycerol/water matrix, the same extract produced a very different spectrum (Figure 3.6(b)). The choice of SDS was based upon the recommendation of Ligon and Dorn, that an excess of a surfactant organic counter ion would increase the surface concentration of the analyte (8).

Another problem which arose shortly after we had obtained success in analyzing standards was the length of operating stability. That is, once a mechanically stable film was obtained on the surface of the probe *in vacuo*, it was not possible to maintain that stability for more than 20-45 minutes.

Consequently, considerable time was spent adjusting all of the instrument operating and focus parameters, which would have to be readjusted when film instability necessitated removing the probe and wiping the surface. The principal cause of instability appeared to be gradual accumulation of glycerol on the probe tip, causing a bubble or drop to form. We reasoned that if the rate of evaporation of solution entering the vacuum chamber were matched by its rate of removal or evaporation, the signal should remain stable for long periods. We have achieved this balance between matrix flow and removal from the probe tip by combining the action of a cellulose wick with increased surface area of evaporation and a larger heat sink (Figure 3.7). This assembly,

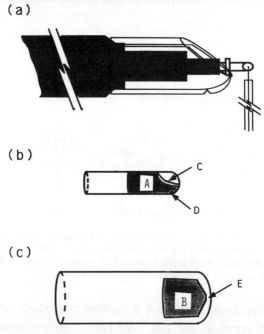

Figure 3.7. (a) Probe evaporator assembly showing a cut-away view of the two sections of Whatman cellulose extraction thimble. (b) Outer layer of the evaporator made from a 10 x 50 mm thimble (shaded portion "A" only is used). (c) The inner layer cut from a 25 x 100 mm thimble (shaded portion "B" only is used by wrapping it to fit inside "A").

fashioned from two layers of cellulose extraction thimble material (Whatman), traps a large percentage of the glycerol which is conveniently removed with the probe at the end of a day of operation. The size of the evaporator used was about 10 mm (i.d.) x 28 mm in length. With 20% glycerol/water mobile phase at 8 µl/min flow rate, the retention of liquid in the evaporator after 9.5 hours of continuous operation was 16.1% by weight. The stability exhibited over 9

Quantitative Analysis

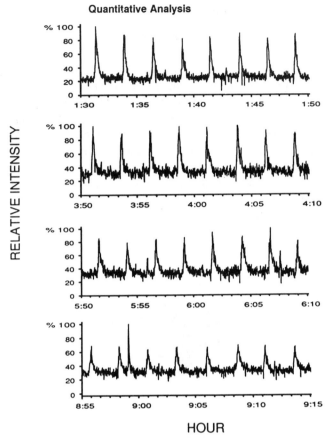

Figure 3.8. Negative ion selected ion recordings of m/z 263 (M-H)⁻ ion obtained by repetitive loop injections containing 10 ng MHPG-sulfate in 0.5 μl of 20% glycerol/water mobile phase extended over more than 9 hours demonstrating the long-term stability obtained with the evaporator device.

hours of operation is shown in Figure 3.8. This evaporator is similar in concept to the wick design used on the Finnigan MAT 90. We have observed that there are at least two additional principles which contribute to long-term film stability. The first is transfer of the liquid from the target surface to a physically separated region for solvent evaporation. This eliminates the requirement for subtle and continuous heat control at the small probe target surface to compensate for the large heat loss due to evaporation, and allows the use of a large heat sink and heat conduction path (i.e., the probe shaft) to provide heat. The second factor is that the use of a relatively large surface area absorbent material in contact with the large heat sink significantly increases the rate of solvent evaporation at reduced temperature.

3.3 FOCUSED CESIUM GUN

We investigated the use of a focused cesium ion gun to maximize the signal-to-noise ratio in order to increase the sensitivity of our quantitative measurements. The work of Aberth *et al.* (9,10) had suggested that cesium LSIMS would be more sensitive than FAB for a variety of compounds, especially with a focused primary ion beam and efficient secondary ion extraction (11). Further, the studies of Barofsky *et al.* (12) suggested that focused liquid metal beams directed at samples deposited on small wire targets significantly enhanced detection of very small quantities of material. We reasoned that the use of an unfocused cesium beam, like a fast atom beam from a saddle field source, would flood the probe, source, and lens regions of the spectrometer as illustrated schematically in Figure 3.9(a). This

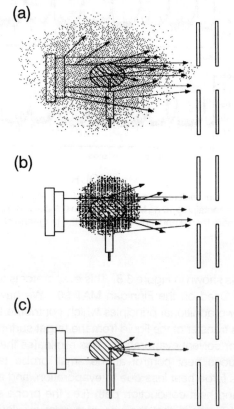

Figure 3.9. Schematic representations of the secondary ion emission obtained with (a) the typical unfocused FAB or cesium ion gun; (b) a focused SIMS gun with a 3 mm diameter beam flooding the target probe of about 1 mm diameter; and (c) a microfocused beam with a diameter significantly smaller than the probe surface area (about 0.1 mm).

phenomenon is well known to instrument operators who have observed the telltale ion burn pattern throughout their ion source, especially on the slit or lens assemblies. A moderately focused beam of about 3 mm diameter would improve conditions in two ways (Figure 3.9(b)). First, it would ensure that only secondary ions generated by the beam centered on the probe tip would be sampled by the spectrometer, thus lowering background chemical noise. Second, the consumption of cesium source elements would be considerably reduced, and the ion source chamber and lens elements would remain free of ion sputter. A microfocused beam (i.e., a beam diameter matched to the secondary ion extraction characteristics of the spectrometer and the probe size) should further improve the analyte signal-to-noise ratio by sampling the eluting liquid stream prior to its dilution across the probe surface, and permit optimization of secondary ion extraction from a point source (Figure 3.9(c)).

The design of a focused cesium ion gun was calculated using computer simulation (Macsimion©), and built employing the unfocused commercial Phrasor cesium gun (3). The gun assembly was retracted from its initial mounting and remounted on an adjustable flange (Acu-port) which permitted correction for alignment and aiming of the beam directly at the probe target (Figure 3.10). The lens voltages for acceleration and optimal focus were close

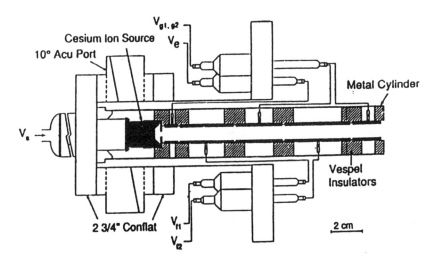

Figure 3.10. Mechanical assembly cross-section of the cesium gun and focusing assembly designed and constructed by T.-C. L. Wang utilizing the commercial (Phrasor Scientific) unfocused gun as a basis. Mechanical adjustability is attained via the 10° motion Acu-Port™ flange.

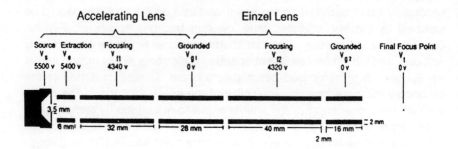

Figure 3.11. Dimensions and voltages used for each cylindrically symmetrical lens element in the focus lens system.

to those calculated (Figure 3.11). The overall length of the assembly was chosen because of the physical constraints of the vacuum chamber and the probe target position. Performance of this gun was first assessed by direct visualization of the beam on sodium iodide-coated metal targets. The primary ion beam appears as an orange spot, whereas stray electrons are seen as blue-white images. Thin-layer ion-sensitive targets are readily prepared in the laboratory by spraying a dilute ethanol solution of sodium iodide (or potassium bromide) onto a warm target surface using an aerosol sprayer, such as a thin-layer chromatography detection reagent sprayer, and air drying the target with a warm heat gun. Alternatively, in order to preserve a permanent image of the beam and accurately measure beam diameter, we used a tantalum pentoxide strip (13). Bentz and Gale noted that refractory metal films or oxides are especially useful when direct visualization of the target is not possible by external viewing, and that such targets can be calibrated by color to yield an accurate measure of integrated beam current. More recently we have employed a probe-mounted Faraday cup detector which has a surface area comparable to the LSIMS target. An electrometer (Keithley 600B) is attached to measure directly the gun current. A daily log of the beam current confirms the consistency of gun performance and assures the operator that primary ions will be correctly directed at the LSIMS target surface. Beam densities between 2 and 5 $\mu A/cm^2$ were found to yield optimal results with an accelerating voltage of 5 kV. As anticipated, the operation of the gun with a focused beam of approximately 3-4 mm diameter produced results comparable to those from the unfocused gun, but without aging of cesium sources over several months of operation, and without any notable accumulation of stray ion burn on the ion source elements. Further, in the microbeam-focused mode, a significant signal-to-noise enhancement was realized, as shown in Figure 3.12. The same sample when analyzed using the

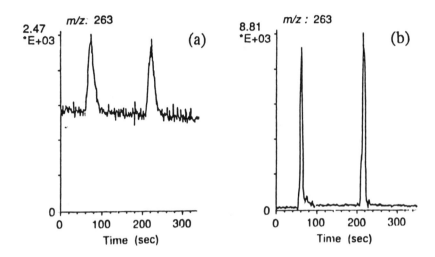

Figure 3.12. Selected ion records obtained when 5 ng MHPG-sulfate was injected (a) with original unfocused commercial ion gun, and (b) with a microfocused beam from the same gun.

unfocused primary ion beam (a) and the focused microbeam (approx. 0.2-0.5 mm (b)) showed the expected beneficial effect due to confining the region of sampling of generated secondary ions to that where the analyte is most concentrated.

Based upon these results, a commercial primary ion gun has been redesigned (Phrasor Scientific) as shown schematically in Figure 3.13. It produces a stable 3 mm diameter beam which floods only the target area to maximize sensitivity as in Figure 3.9(b). We measured a beam current of 1-5 $\mu A/cm^2$ at 5 kV using a Faraday cup probe with a target the same size (approximately 1 mm^2) as the conventional LSIMS probe tip. The integrated secondary ion signal from repetitive injections of a convenient test compound, 25 ng aliquots of 4-sulfatoxy-3-methoxyphenylethylene glycol (MHPG-sulfate), were plotted as a function of heater setting and the measured beam current (Table 3.1). As reported by others, the secondary ion signal increases nearly linearly with increasing beam current until the beam density causes greater damage than useful analyte ion generation at approximately 4-5 $\mu A/cm^2$. The

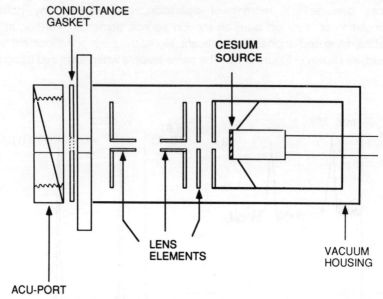

Figure 3.13. Schematic representation of the simplified and optimized focusing lens system of a prototype Phrasor Scientific cesium gun producing a stable 3 mm beam which can be mechanically steered to align with the probe target.

best measure of useful secondary ion current is probably sample dependent, as the optimum for matrix and analyte are likely to differ, making the common practice of tuning on matrix ions less than ideal.

Table 3.1. Secondary Ion Signal as a Function of Beam Current

Heater Setting	Beam Current (μA/cm^2)	Peak Area[a] x 10^{-4}
1.75	0.82	21.6
2.00	1.7	48.2
2.25	2.6	71.5
2.50	3.6	97.1
2.75	4.2	71.6
3.00	5.3	82.1

[a] Peak areas are averages of three measurements of integrated (M-H)$^-$ intensity for 25 ng injections of MHPG-sulfate at m/z 263 in CF/LSIMS mode.

3.4 CONCLUSIONS

We have shown that the rapid and routine quantitative analyses of those polar organic compounds which ionize well by FAB or LSIMS is very reasonable, with the caveats which accompany all emerging techniques. In our experience, the major challenge for quantitation of compounds from biological sources lies in coupling suitable sample concentration and separation technology with CF-LSIMS to avoid signal suppression. Either off-line separation processes, on-line microbore HPLC, or on-line HPLC with liquid concentration are attractive approaches depending upon analyte concentration and the required sensitivity of analysis.

ACKNOWLEDGMENTS

The authors are indebted to Peter L. Todd of Oak Ridge National Laboratory for his advice and assistance in ion gun design and beam current measurements. The Farady cup probe used in our studies was fabricated by Peter L. Todd. The efforts of Tao-chin Lin Wang, M. Duncan, R. Boni, and Louis Cornio have significantly enhanced this research.

REFERENCES

1. T.-C.L. Wang, M.-C. Shih, S.P. Markey, and M.W. Duncan, *Anal. Chem. 61*, 1013 (1989).
2. M.-C. Shih, T.-C.L. Wang and S.P. Markey, *Anal. Chem. 61*, 2582 (1989).
3. T.-C.L. Wang, S.P. Markey, and R.L. Boni, *Proc. 37th ASMS Conf. Mass Spectrom. and Allied Topics*, Miami Beach, FL, 1989, p.1039.
4. D.F. Hunt, J.R. Yates,III, J. Shabanowitz, S. Winston, and C.R. Hauer, *Proc. Natl. Acad. Sci. USA 83,* 6233 (1986).
5. M.E. Hemling, S.A. Carr, C. Capiau, and J. Petre, *Biochem. 27,* 699 (1988).
6. D.F. Hunt, J.R. Yates, III, J. Shabanowitz, M.E. Bruns, and D.E. Bruns, *J. Biol. Chem. 264,* 6580 (1989).
7. D.-C. Moon, and J. A. Kelley, *Biomed. Environ. Mass Spectrom. 17,* 229 (1988).
8. W.V. Ligon, and S.B. Dorn, *Int. J. Mass Spectrom. Ion Processes 61,* 113 (1984).
9. W. Aberth, K. Straub, and A.L. Burlingame, *Anal. Chem. 54,* 2029 (1982).
10. W.H. Aberth, and A.L. Burlingame, *Anal. Chem. 60,* 1426 (1988).
11. W. Aberth, *Proc. 32nd ASMS Conf. Mass Spectrom. and Allied Topics*, San Antonio, TX, 1984, p.604.
12. B. Arbogast, E. Barofsky, L.F. Jiang, and D.F. Barofsky, *Proc. 37th ASMS Conf. Mass Spectrom. and Allied Topics*, Miami Beach, FL, 1989, p.1151.
13. B.L. Bentz, and P.J. Gale, *Anal. Chem. 55,* 1434 (1983).

Chapter 4

DIRECT ANALYSIS OF BIOLOGICAL PROCESSES

Richard M. Caprioli

One of the major advantages that CF-FAB provides to the analyst is the ability to directly analyze aqueous solutions with mass spectrometry (1, 2). This capability is of importance because, generally, biological samples are contained in water solutions and most dissolved analytes are polar or charged molecules. CF-FAB can be used to analyze such molecules without derivatization by permitting the direct injection of aqueous solutions without purification or concentration. In addition, the technique typically shows sensitivity enhancements of 10-100-fold or more over standard FAB (3).

CF-FAB mass spectrometry can be used for batch sample processing in a number of different operating modes (4). First, it can be employed for constant flow reaction monitoring, i.e., where one or more compounds in a chemical or biochemical reaction is followed continuously. This is most advantageous when reactant or product concentrations change rapidly or where it is inconvenient or inadvisable to take individual sample aliquots. Second, it can be used in a flow injection analysis mode whereby a carrier solution is constantly allowed to flow into the mass spectrometer source and samples are injected into this carrier flow. This mode has the advantage of allowing high-concentration injections over very short periods of time, giving high signal-to-noise ratios. It also produces a temporal relationship between the background of the carrier solution and the ions obtained from the sample injection and, thus, provides effective background subtraction. A typical use may be, for example, the analysis of a batch reaction mixture where sample aliquots can be taken at time intervals and injected, either manually or automatically, as the reaction proceeds. This flow injection mode is also advantageous for the processing of a large number of samples for the routine analysis of one or more compounds. Under optimal conditions, we have found that sample injections can be made at a rate of about one sample every 2 minutes with excellent results.

This chapter will be devoted to applications of the use of CF-FAB for biological samples using both constant flow and flow injection analysis modes.

Continuous-flow Fast Atom Bombardment Mass Spectrometry
Edited by R.M. Caprioli © 1990 John Wiley & Sons Ltd

The applications will include examples of batch sample processing, enzyme reaction analysis in real-time, kinetic analysis of enzyme activity and microdialysis/MS for *in vivo* drug monitoring and on-line enzyme reaction monitoring.

4.1 FLOW INJECTION ANALYSIS

Flow injection analysis is the most advantageous method of operation of CF-FAB because it provides the analyst with a great deal of versatility in the analysis of different types of samples under different conditions. Typically, a carrier solution which contains 95% water and 5% glycerol is allowed to flow into the mass spectrometer source at a rate of approximately 5-10 μl/min. Additives such as trifluoroacetic acid, acetonitrile, salts, etc., can be used to enhance ion production and are used at low concentrations. Figure 4.1

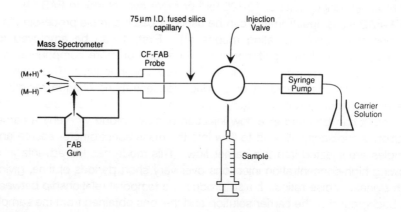

Figure 4.1. Instrumental arrangement for flow injection analysis for CF-FAB mass spectrometry. (Reprinted with permission from reference 17.)

illustrates the instrumental set-up for flow injection analysis. The sample is injected into the carrier flow via a microinjector valve. Sample injection volumes of 0.5 -5 μl are generally used, although a 1 μl injection is a good trade-off between handling convenience and sample exposure time. For example, a 1 μl sample will be exposed to the atom beam for approximately 30-40 seconds with a 5 μl/min carrier flow rate.

Ions from a sample that is flow-injected are produced in a specific time domain and appear as a single "chromatographic" peak in the total ion chromatogram. This is shown in Figure 4.2(a) for the compound substance P, a peptide having a molecular weight of 1347. The mass spectrum is shown in Figure 4.2(b).

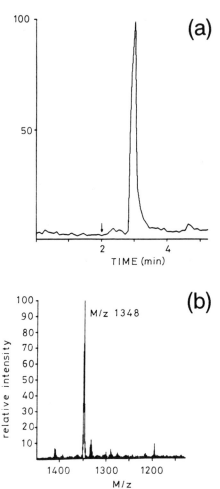

Figure 4.2. Analysis of substance P by CF-FAB using flow injection techniques. (a) Injection peak profile of 13.5 ng in 0.5 μl injected, and (b) mass spectrum of $(M+H)^+$ region from this sample injection.

CF-FAB Mass Spectrometry

Using software generally present on MS data systems for GC or LC chromatographic analyses, mass spectra can be obtained substantially free of background by subtracting spectra that are adjacent to the "chromatographic" sample peak. This is illustrated for the analysis of a peptide of $(M+H)^+$ 845 as shown in Figure 4.3. Figure 4.3(a) shows the standard FAB mass

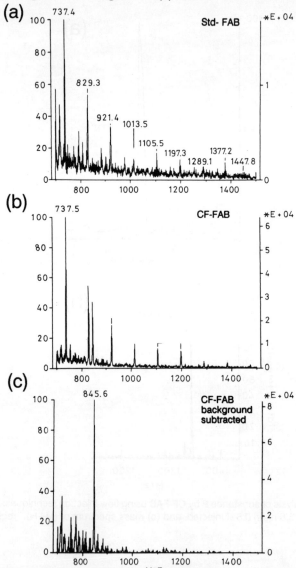

Figure 4.3. Mass spectra from the analysis of a synthetic peptide (844 daltons) with (a) standard FAB, (b) CF-FAB, and (c) CF-FAB with background subtracted.

spectrum; (b) the spectrum obtained by flow injection analysis using CF-FAB without background subtraction; (c) the spectrum obtained with CF-FAB after subtraction of the background preceding the sample peak. It is clear that for the standard FAB spectrum, there is no discernible molecular ion species and, therefore, no further meaningful data work-up is possible. For the spectrum obtained with CF-FAB, a molecular ion species is detectable and this ion is significantly enhanced by background subtraction.

A second major advantage of the flow injection mode of analysis is that it allows quantitative comparison of sample injections without the use of internal standards. After the CF-FAB probe is inserted into the mass spectrometer, it should remain untouched throughout a series of analyses so that the position and orientation of the target is constant. Quantitative comparisons of ion intensities are then possible, with calculated areas under (injected) peaks in ion chromatograms showing deviations from the mean of less than ±10%. Quantitative measurements of better than ±5% can be achieved and this is discussed in some detail in Chapter 3. The ability to quantitate without internal standards is a great asset, particularly in light of the fact that reproducible analysis with standard FAB is extremely difficult because the placement and shape of the liquid sample droplet on the FAB target and the precise positioning of the probe itself in a highly reproducible manner is nearly impossible. For standard FAB, these problems produce extreme variations in ion intensities and the inability to perform quantitative analyses from sample to sample without the use of internal standards.

Several microinjector valves connected in series can be used as a simple liquid sample interface for a mass spectrometer, as shown in Figure 4.4. Valve 1, a Rheodyne 7010 or equivalent, is used as a bypass to allow lines from the pump to be quickly cleaned and flushed when changing carrier solutions. Valve 2, a Rheodyne 7410 or equivalent, is a four-port valve that can be used to directly sample reaction solutions. One of the ports can be connected to a vacuum source which is turned on and off by an electrically operated solenoid valve. Thus, at appropriate times, aliquots of sample solution from the reaction can be loaded into the injection loop and subsequently flow injected for the analysis. Note that for this application, valves 1 and 3 are in the bypass position. It is convenient to operate valve 2 electrically so that a microprocessor can automate the entire procedure including sample loading and injection. Valve 3, a Rheodyne 8125 or equivalent, is used for direct sample injection. It is recommended that this valve be a "no dead-volume" type which allows the syringe needle to be placed directly against the stator so that there is no sample waste in filling the valve. Thus, a 1 μl injection volume requires 1 μl delivered from the sample syringe. For valves not designed in this fashion, it is not unusual to have to inject 5 to 6 μl of solution for a 1 μl injection loop to insure that the valve is flushed and

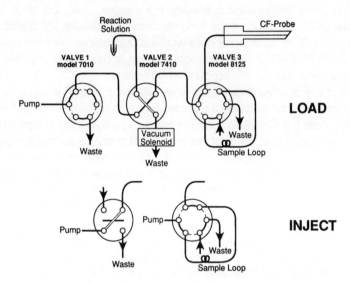

Figure 4.4. Arrangement of microvalves for a simple liquid/MS interface. The top portion of the figure shows the valves in the load position, i.e., sample could be loaded in valves 2 or 3, and the bottom portion shows sample valves 2 and 3 in the "injection" position.

the sample loop is fully loaded. Valve 3 also makes it convenient to inject a standard for purposes of either mass calibration or instrument tuning. Other valves can also be added for sample prepurification or continuous sample collection procedures if desired.

Great care must be taken in making all plumbing connections to eliminate dead volumes. Only ferrules designed specifically for a given injector must be used. For connections involving fused silica capillaries, fittings supplied by Upchurch Scientific for specific valves give excellent dead-volume free connections. Also, capillaries must not protrude beyond the end of these ferrules, since they can enter the valve and damage the stator.

4.2 BATCH SAMPLE PROCESSING

Flow injection analysis using CF-FAB is an extremely effective method for analyzing large numbers of samples. Under the proper conditions, a sample may be injected once every 2 minutes. One of the concerns in such an application is the memory effect caused by the presence of a compound from the "tail" of the previous injection. To test the capability of the system, Figure 4.5 shows the results of an experiment in which two peptides, substance P

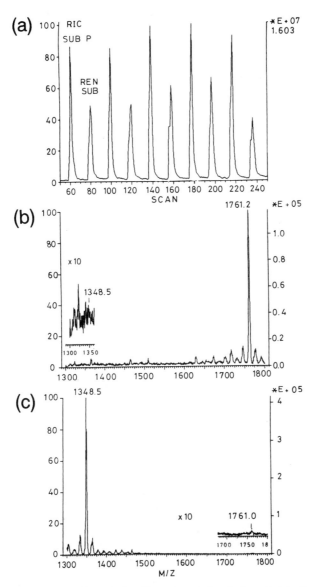

Figure 4.5. Mass spectral data obtained from the continual alternate injection of 200 pmol of substance P and renin substrate tetradecapeptide. (a) Total ion chromatogram showing the first ten injections; (b) the mass spectra of the last (20th) injection of renin substrate; and (c) the last (20th) injection of substance P. See text for details.

$[(M+H)^+ = m/z$ 1348] and renin substrate tetradecapeptide $[(M+H)^+ = m/z$ 1720] were alternately injected at a level of 200 pmol/μl every two minutes over a period of approximately 1.3 hours. Figure 4.5(a) shows a portion of the total ion chromatogram from this series of injections. Figure 4.5(b) shows the last injection of substance P with the area around the molecular species of renin substrate expanded in intensity by a factor of ten. Figure 4.5(c) shows the last injection of renin substrate with the area around the molecular species of substance P expanded by a factor of ten. It is seen that under these conditions very little, if any, carryover is seen from injection to injection. Note that the peak widths in time at baseline are approximately 50 seconds using 1 μl sample injections at a flow rate of approximately 5 μl/min. It is emphasized that for rapid sample injection rates such as this, one must take care to eliminate all dead volumes in the plumbing connections to minimize peak tailing.

The use of MS/MS may be extremely important, particularly for the direct analysis of biological fluids or other chemically complex samples, because of its ability to reduce background and interference between ions. In a triple quadrupole MS/MS system (Figure 4.6), sample ions produced in the source

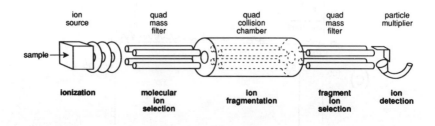

Figure 4.6. Diagram of a triple-quadrupole MS/MS (tandem) mass spectrometer.

from the bombardment process are individually selected by the first quadrupole mass filter. Thus for a particular $(M+H)^+$, these ions are transmitted into the second quadrupole, operated to pass all ions without mass selection. In this second quadrupole, the ion selected by the first quadrupole is allowed to collide with a gas molecule, causing decomposition of this parent ion to form daughter ions. The third quadrupole is then used as a mass filter to scan through the mass spectrum and record the masses of the individual daughter ions produced in the collisionally activated decomposition process. The net result is a mass spectrum of fragment ions produced from the single $(M+H)^+$ parent ion even though this parent ion was originally produced in a complex mixture of ions. This MS/MS technique therefore eliminates much of the background ions normally produced by desorption ionization. However, it is noted that at any one *m/z* value, several ionic species may occur and, at nominal resolution, all of these ions would be passed into the collision region.

An example of batch sample processing using CF-FAB combined with MS/MS is the work of Seifert *et al.* (5) in the analysis of drugs in urine. Drugs analyzed in this work included benzylecognine, cocaine, codeine, morphine, penicillin G, pentobarbital, phenobarbital, phentolamine, and delta-9-tetrahydrocannabinol. The instrument was first calibrated and standard responses recorded, as shown in Figure 4.7, for the analysis of known

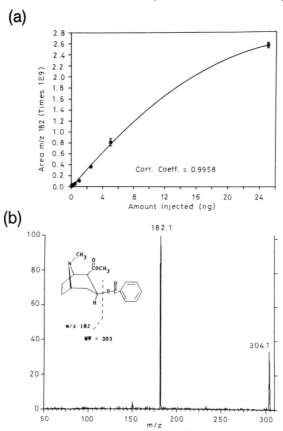

Figure 4.7. Analysis of cocaine in aqueous solution with CF-FAB. (a) Standard curve using the daughter ion at *m/z* 182 for quantitation; (b) the daughter ion mass spectrum.

concentrations of cocaine. Figure 4.7(a) shows the standard curve obtained by plotting ion intensity from the major fragment ion in the MS/MS spectrum (*m/z* 182) versus the concentration of drug. A nonlinear least squares fit of the data shows a correlation coefficient of 0.9958 with a linear portion between 0.1 and 10 ng of cocaine injected. A sample MS/MS spectrum is shown in Figure 4.7(b) where a small amount of parent ion at *m/z* 304 can be seen, which has survived the collision process, together with the major fragment ion at *m/z* 182.1. Figure 4.8(a) shows the multiple injection profile for the selected ion chromatogram for *m/z* 182+304 of a urine sample containing 10 ng/µl of cocaine and Figure 4.8(b) the MS/MS mass spectrum from one of these injections.

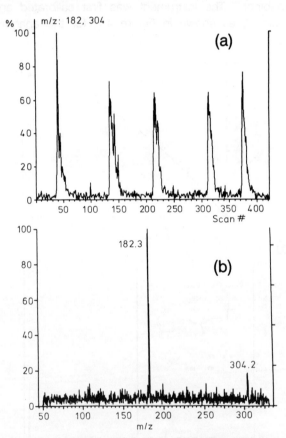

Figure 4.8. Direct analysis of cocaine in human urine with CF-FAB. (a) The total ion chromatogram for five replicate injections of 10 ng in 1 µl of urine; (b) the MS/MS daughter ion spectrum of one of these injections. (Reprinted with permission from reference 5.)

4.3 ON-LINE REACTION MONITORING

On-line reaction monitoring involves the recording of the ion intensity of one or more reactants or products over a period of time with samples taken directly from an on-going reaction. Both the flow injection mode and the constant flow mode of operation of the CF-FAB interface can be used effectively in such an application (4,6). For reactions in which the concentrations of analytes are changing relatively slowly, flow injection analysis is advantageous because it maximizes sensitivity and allows efficient background subtraction. Typically, sample aliquots can be taken and analyzed every 2-3 minutes. When analyte concentrations are changing relatively rapidly, the constant flow mode has greater advantage because there is no dead time between analyses other than instrument scan time. A consequence of the constant flow technique is that either the reaction mixture must be directly compatible with the FAB ionization processes or a suitable matrix solution must be added after the sample is withdrawn through the use of a mixing tee or other device. This last procedure, although effective, has the disadvantage of diluting the sample solution.

4.3.1 Analysis of Aliquots by Flow Injection

An example of reaction monitoring utilizing flow injection analysis is the measurement of kinetic constants for the reaction of the enzyme trypsin with the polypeptide substrate Met-Arg-Phe-Ala (4). Figure 4.9(a) shows the automated reaction analysis for the action of a constant amount of enzyme on five different concentrations of the substrate; 1.25 mM, 1.0 mM, 0.75 mM, 0.5 mM and 0.25 mM. The areas under the peaks for each individual concentration were then plotted and the rate for that particular reaction was calculated. Figure 4.9(b) shows the plot of the individual rates at the five substrate concentrations in a Lineweaver-Burke plot of $1/V$ versus $1/S$, where V is the velocity or rate of the reaction and S is the substrate concentration. This double-reciprocal plot has an x-axis intercept of $-1/K_M$, giving a value for the Michaelis constant, K_M, of 1.85 mM, and a y-axis intercept of $1/V_{max}$, with a calculated value of V_{max} (the maximum velocity) of 0.11 mM/min. These values are in excellent agreement with a previously published value for the $K_M = 1.90$ (7).

A second application which shows the importance of the mass specific information obtained by mass spectrometry, even in very complex mixtures, is the reaction of a single enzyme with a multiple substrate reaction mixture. Table 4.1 lists the sequence and the value of the $(M+H)^+$ for each of the reactant and product ions expected from the action of trypsin on a mixture of

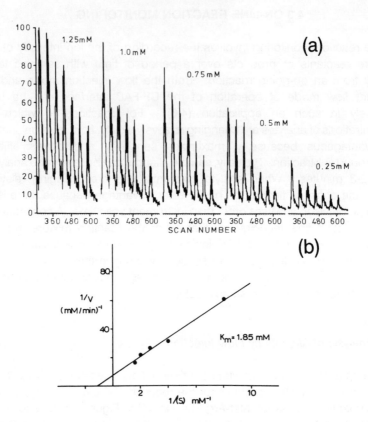

Figure 4.9. Automated flow injection analysis of the tryptic digest reaction mixture for the hydrolysis of the peptide Met-Arg-Phe-Ala. (a) Selected ion recording of the molecular species (m/z 524) for five substrate concentrations (left to right), 1.25, 1.0, 0.75, 0.5, and 0.25 mM, and (b) double-reciprocal plot of the rate data obtained from measurements of areas under peaks in panel (a). (Reprinted with permission from reference 4.)

eleven different substrates. Since the mass spectrometer is a multi-dimensional analyzer, individual reactants and products can be followed independently as long as their $(M+H)^+$ values are different, assuming the instrument is set at unit resolution. Quantitative measurements of the areas under the injected peak profiles versus time for each $(M+H)^+$ of interest provide the required kinetic data. For example, Figure 4.10(a) shows the decrease in the molecular ion species of physalaemin at an $(M+H)^+$ of 1265 and the concomitant increase in that of the product ion at m/z 629. Figure 4.10(b) shows the rate plots obtained from each individual injection peak profile for three different substrates, illustrating the range of the reaction rates observed in one such experiment.

Table 4.1. Composition of Peptide Mixture Used for On-line Monitoring of Trypsin Digestion

Peptide	Sequence	[M+H]$^+$ (m/z) Reactant	Products
Renin Substrate	Asp-Arg-Val-Tyr-Ile-His-Pro-Phe-His-Leu-Val-Ile-His-Asn	1759	1488, 290
Neurotensin	pGlu-Leu-Tyr-Glu-Asn-Lys-Pro-Arg-Arg-Pro-Tyr-Ile-Leu	1672	1030, 661
α-MSH	Ac-Ser-Tyr-Ser-Met-Glu-Phe-His-Arg-Trp-Gly-Lys-Pro-Val(NH$_2$)	1665	1098, 585
Bombesin	pGlu-Gln-Arg-Leu-Gly-Asn-Gln-Trp-Ala-Val-Gly-His-Leu-Met(NH$_2$)	1620	1224, 414
Angiotensin I	Asp-Arg-Val-Tyr-Ile-His-Pro-Phe-His-Leu	1296	1025, 290
Physalaemin	pGlu-Ala-Asp-Pro-Asn-Lys-Phe-Tyr-Gly-Leu-Met(NH$_2$)	1265	629, 584
Angiotensin II	Asp-Arg-Val-Tyr-Ile-His-Pro-Phe	1046	775, 290
Angiotensin III	Arg-Val-Tyr-Ile-His-Pro-Phe	931	775, 290
Met-Enkephalin-RF	Tyr-Gly-Gly-Phe-Met-Arg-Phe	877	730, 290
Gly Octapeptide	Gly-Arg-Gly-Leu-Ser-Leu-Ser-Arg	845	632, 232
Kemptide	Leu-Arg-Arg-Ala-Ser-Leu-Gly	773	503, 288

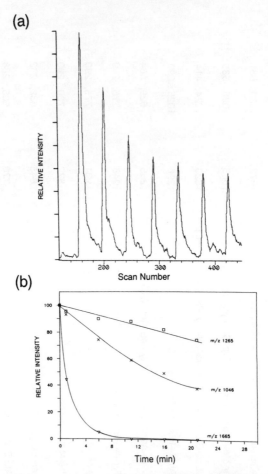

Figure 4.10. On-line reaction monitoring of the tryptic hydrolysis of a peptide substrate "cocktail" (listed in Table 4.1). (a) Selected ion chromatogram for the automated flow injection analysis of one substrate, physalaemin ([M+H]$^+$ = 1265), and (b) rate data for the hydrolysis of three substrates.

One can also monitor the action of multiple enzyme mixtures on a single-reaction substrate. In such an experiment, the endopeptidase subtilisin was mixed with several exopeptidases (carboxypeptidase Y, A and B) and the rate of production of the various fragments produced were followed in a time-course reaction (4). In the case of the hydrolysis of the peptide bombesin (1620 daltons), the exopeptidases were present at approximately a ten-fold excess activity over the subtilisin. The endopeptidase produced two major fragments, m/z 825 and m/z 812, which were observed within a few minutes of starting the reaction. The carboxypeptidases hydrolyzed the C-terminal residues from the intact polypeptide, as well as the two fragment polypeptides, producing a series of molecular species. From these ions, a string of masses could be identified which were related to each other by the mass of a single amino acid residue, as shown in Figure 4.11. All of the molecular species down to the tetrapeptide were observed. It should be noted that even if one or two peptides in the string are not observed, this would not prevent re-establishment of the string at lower mass and subsequent sequence determination in this range.

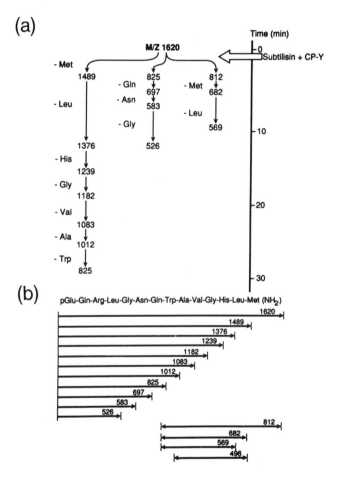

Figure 4.11. Time-course production of peptides identified by CF-FAB mass spectrometry from the on-line hydrolysis of bombesin with a 10:1 mixture of carboxypeptidase Y:subtilisin. (a) The approximate time of appearance of the various fragments, and (b) the arrangement of these ions in a string where they are related by the mass difference of an amino acid residue. (Reprinted with permission from reference 4.)

The effective use of the time domain in experiments that follow enzyme hydrolyses can provide a great deal of additional information on intermediate structures. This is illustrated by the tryptic hydrolysis of the peptide calcitonin followed over a period of approximately 30 minutes (8). Since not all tryptic cleavages occur at the same time, a series of peptides are produced during the course of the reaction which, in fact, represent peptide overlaps that can be used for the purposes of aligning the tryptic fragments that remain at the end of the reaction. Relatively low concentrations of enzymes are used to allow intermediates to be seen, i.e., enzyme/substrate molar ratio of 1:1000. For calcitonin from eel, analysis of the final reaction products of a tryptic digest, as typically done for a FAB mapping experiment, showed four polypeptides at m/z 1123, 855, 778 and 717. Of course, their order relative to each other in the original molecule cannot be determined from these data. Analysis of multiple time points during early phase of the reaction showed the presence of five intermediates at m/z 2719, 2312, 1961, 1614, and 1476. Since these peptides represent sequence overlaps, it is possible to correlate the position of each peptide produced in a single clip relative to its neighbor. The individual intermediates identified in this way are shown in Figure 4.12. In the

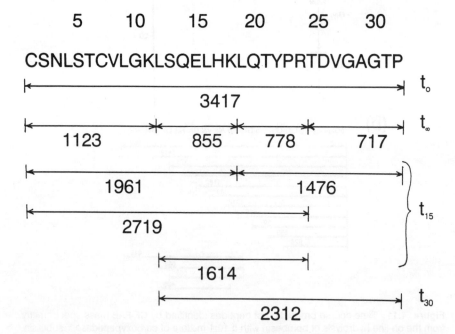

Figure 4.12. Peptide products identified at various times during the reaction of trypsin with eel calcitonin. The enzyme:substrate molar ratio was 1:2000. The numbers below the lines represent the $(M+H)^+$ values measured for the peptides.

mass spectrum, two ions produced from a common precursor peptide are easily identified because the sum of the $(M+H)^+$ values of the two fragment peptides are equal to the $(M+H)^+$ value of the precursor peptide minus 19 mass units (H_3O^+). With a knowledge of the C-terminal or N-terminal residue to establish direction, a unique "sequence ordered" tryptic fragment map can be produced from a single time-course analysis. It should be noted that this kind of data analysis is not a kinetic function where time must be measured accurately and incorporated into calculations; time merely serves to orient peptides relative to others.

4.3.2. Constant Flow Analysis

The continuous or constant flow of an on-going reaction mixture into the source of the mass spectrometer through the CF-FAB interface provides an uninterrupted and mechanically simple method of reaction monitoring (4, 6). In many cases, this can be done directly from an open-reaction vial by placing the end of the fused silica capillary directly in the reaction solution. This technique requires that the reaction solution is compatible with the FAB ionization process, i.e., that pH, solvents, salts, matrix additive, etc. present in the reaction allows the formation of adequate ion intensities. The instrumental set-up for such an experiment is shown in Figure 4.13 where atmospheric

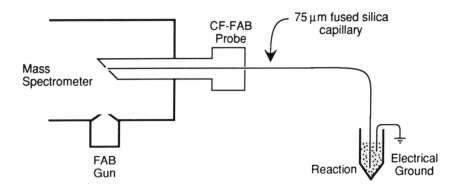

Figure 4.13. Instrumental arrangement for constant reaction monitoring with CF-FAB using atmospheric pressure to produce liquid flow. (Reprinted with permission from reference 19.)

pressure is used to drive the reaction solution through the capillary and into the mass spectrometer. This arrangement utilizes a constant pressure, not a constant flow rate, and caution should be taken to insure that there is no particulate matter in the system that could block the capillary and alter the flow rate and stability of the system. In addition to providing a nearly continual output of data, it has the obvious advantage of eliminating mechanical pumps and dead-volume connections.

An example of constant flow reaction monitoring is provided by following the enzymic reaction for the hydrolysis of ribonuclease S peptide by a combination of carboxypeptidases Y and P (2). The reaction solution was constantly flowing into the ion source at a rate of 4 μl/min and mass spectra continuously taken throughout the reaction. Glycerol was not present in the reaction but was added in the CF-FAB interface in a low-dead-volume tee connection (1:1 split) using a glycerol/water (3:7) solution containing 0.3% TFA. The enzymic reaction contained approximately 100 pmol/μl of substrate in 50 mM Tris pH 6.0 buffer. Figure 4.14(a) shows four of the molecular species produced as a function of time. The ion at m/z 2095 represents the $(M+H)^+$ for the original polypeptide; m/z 2024, the truncated peptide formed from the loss of the C-terminal residue; m/z 1937, the loss of four C-terminal residues; and m/z 1547, the loss of six C-terminal residues. In all, fourteen molecular species were identified, corresponding to the successive loss of 13 amino acids from the C-terminal of the original polypeptide. One of the advantages of this approach in following the truncated molecular species for the polypeptide, and not the residues released, is that sequence information based on the mass difference of the ions, i.e., molecular species are produced in a sequence and are separated by a mass equal to that of an amino acid residue. Of course, a time shift in the intensity maximum of these ions can be seen and these data can be of value in cases where ambiguities arise, particularly when substrates are present at low concentrations. A single mass spectrum from this hydrolysis is shown in Figure 4.14(b) for the reaction mixture at 20 minutes, approximately at scan 150. Most of the molecular species are relatively intense and are labeled in the spectrum, with the less intense peaks labeled "a" being sodium adduct ions, and "b" and "c" decomposition ions caused by the FAB process.

4.4 MICRODIALYSIS

Dialysis is a technique based on the ability of a membrane to pass a solute from an area of higher concentration to that of lower concentration. It is an equilibrium process and thus, in a simple system, eventually both solutions will be at the same concentration. Microdialysis involves the use of a small probe

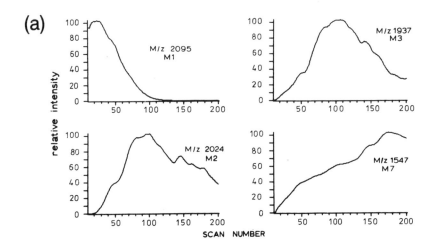

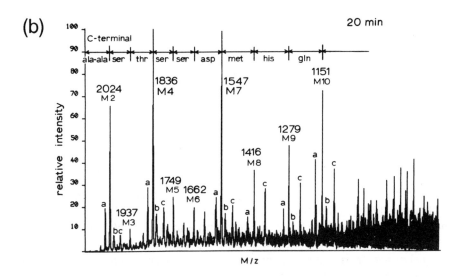

Figure 4.14. On-line monitoring of the hydrolysis of ribonuclease S peptide by carboxypeptidases Y and P. (a) Selected ion chromatograms for several truncated molecular species, and (b) a single mass spectrum taken 20 minutes into the reaction. Peaks labeled "a" are sodium adduct ions, and "b" and "c" are artefacts of the bombardment process. Additional details are given in the text. (Reprinted with permission from reference 2.)

containing a dialysis membrane which can be placed into solutions and biological tissues to sample dialyzable compounds in these systems. Indeed, such a device, shown in Figure 4.15, has been used for the analysis of

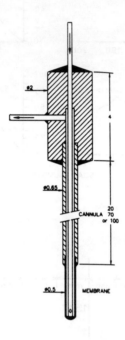

Figure 4.15. Microdialysis probe manufactured by Carnegie Medicin, Uppsula, Sweden. (Courtesy Carnegie Medicin.)

neurotransmitters in brain tissue of live animals using electrochemical detection techniques (9-11). The device, supplied commercially by Carnegie Medicin, has a polycarbonate membrane 3-4 mm long with an area of approximately 5 mm^2 and has a molecular weight cutoff of approximately 20,000 daltons. In normal operation, an aqueous carrier solution is passed inside the probe at a rate of 2 - 5 μl/min to sweep out the membrane reservoir into which analytes have dialyzed. Thus, molecules from outside the membrane will diffuse into the inside reservoir and are subsequently swept out of the dialysis probe to be analyzed. Mass spectrometry has been coupled with microdialysis through the CF-FAB interface for specific mass detection of the dialysate (12).

4.4.1. Characteristics of Microdialysis

The operation and capabilities of the microdialysis probe are critical in these applications and will be briefly reviewed using documentation supplied by the manufacturer (13). Figure 4.16 shows a comparison of the flow rate versus

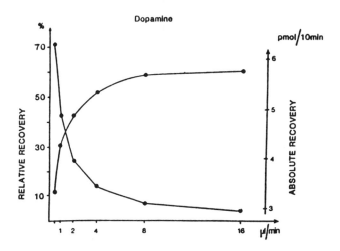

Figure 4.16. Relationship between perfusion flow rate, and relative and absolute recovery with a microdialysis probe. (Reprinted with permission from reference 13.)

relative recovery and absolute recovery for dopamine in aqueous solutions. With respect to relative recovery, the higher the rate of perfusion of the inside reservoir of the microdialysis probe, the lower is the relative recovery. Thus, the concentration of analyte in the dialysate decreases with higher flow rates. However, the absolute recovery, defined as the amount of analyte recovered in a given period of time, first increases and then is constant with higher flow rates because the dialysis is driven at a maximal net rate due to the efficient removal of analyte in the inner reservoir. Table 4.2 lists the relative recoveries *in vitro* for a number of substances for 3 mm and 4 mm lengths of membrane using a flow rate of 2 μl/min. It can be seen that for the 3 mm membrane, average recoveries of organic molecules are in the range of 15-20%.

Another important operating parameter of the microdialysis probe is its response time to concentration changes (14). This was tested using a microdialysis probe with a 4 mm long membrane and a perfusion rate of 2 μl/min. The concentration of four glucose solutions containing 0.13%, 0.3%, 0.5% and 0.13% glucose were analyzed by consecutively shifting the probe

Table 4.2. Relative Recoveries of Substances with Microdialysis at 2 µl/min Flow Rate (from reference 13.)

Substance	(%) Relative Recovery	
	3 mm	4 mm
Dopamine	19	24
DOPAC	19	24
5-HT	17	24
5HIAA	17	24
HVA	19	24
Noradrenaline	18	23
NA+	39	49
K+	61	71
Pyruvate	38	45
Lactate	34	39
Hypoxanthine	23	31
Inosine	17	21
Guanosine	16	21
Adenine	21	29
Adenosine	14	19
Glucose	18	22
Acetylcholine	24	27
Choline	26	31
Aspartate	16	24
Asparagine	17	25
Glutamate	16	24
Serine	14	22
Glutamine	16	24
Taurine	22	27
Tyrosine	16	20
GABA	21	25
Tryptophan	19	27
Methionine	18	27
Valine	18	26
Phenylalanine	18	26
Isoleucine	18	26
Leucine	19	26

from sample to sample. Each solution was dialyzed for 5 minutes and 2 µl aliquots were collected for analysis. Figure 4.17 shows the results of this experiment with the arrows indicating the points at which the microdialysis probe was shifted to the next solution. Overall, the results showed that the delay in response to concentration change in this system is approximately 2 to 3 minutes. This will, of course, be dependent upon the chemical and physical character of the compound of interest and physical characteristics of the experimental set-up.

The effect of temperature on the dialysis process was also measured for a glucose solution using a 3 mm probe at a perfusion rate of 1.5 µl per minute

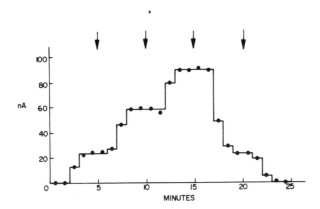

Figure 4.17. Delay associated with microdialysis resulting from switching the probe between four solutions containing glucose at concentrations of 0.13%, 0.3%, 0.5%, and 0.13%. Each solution was dialyzed 5 minutes and 2 μl samples collected and analyzed. The arrows mark the time of sample switching. (Reprinted with permission from reference 14.)

(14). The results show a gradual increase in the percentage relative recovery as temperature increases with a recovery of approximately 12% at 0 °C, 18% at 25 °C and 24% at 40 °C. The mid-point in this curve is centered at about 25 ° and is relatively flat without a significant change in recovery in a ±5 °C window.

Although binding of substances to the membrane has not appeared to be a problem for the substances shown in Table 4.2, the possible loss of material in this way should be kept in mind. However, because of the small surface area of the membrane, binding of the analyte is more likely to occur in the walls of the tubing and valves used to transfer the dialysate.

4.4.2 In Vivo Drug Monitoring

In preparation for measuring drug levels on-line in live animals, we studied the efficiency of the microdialysis/MS apparatus in following concentration changes in an experimental apparatus (15). This apparatus consisted of a solution of the antibiotic penicillin G at a concentration of 100 ng/μl in a vial which was efficiently stirred and which contained the microdialysis probe. The perfusate from the probe was led to a microinjector valve having two sample loops so that each sample loop could be alternately filled and analyzed by the mass spectrometer every 10 minutes. The concentration of drug in the reservoir was changed by either the addition of water or a concentrated solution of the drug. The results are shown in Figure 4.18 where the solid line represents the

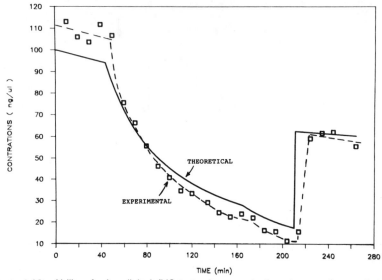

Figure 4.18. Ability of microdialysis/MS to track concentration changes for a solution of penicillin G by dilution and concentration of drug. The solid line represents the theoretical (calculated) concentration and the broken line the experimentally determined concentration.

calculated (theoretical) concentration change (complete and instantaneous mixing was assumed) and the broken line, the experimental measurements. It is seen from this curve that under the conditions specified, the data obtained from the microdialysis/MS combination accurately follow the concentration changes occurring in the reaction vessel.

The experimental set-up for the *in vivo* pharmacokinetic analysis of penicillin G in the rabbit is shown in Figure 4.19 as described by Caprioli and Lin (15). In this experiment, 150 mg of the drug was infused into the ear vein of a rabbit over a 20-minute period. The microdialysis probe was inserted into the jugular vein and the perfusate outlet tube connected to a microinjection valve containing two 10 μl sample loops. The contents of the loops could then be emptied directly into the mass spectrometer through the CF-FAB interface. The procedure for the insertion of the microdialysis probe into the jugular vein of the rabbit is depicted in Figure 4.20. It is noted that animal care guidelines from the National Institutes of Health and from the University Animal Welfare Committee were rigorously adhered to. In this study, MS/MS techniques were utilized to reduce background and interfering ions due to the chemical

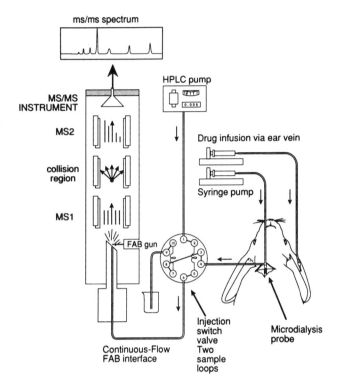

Figure 4.19. Instrumental set-up for *in vivo* penicillin G pharmacokinetic study in the rabbit. The drug was infused in the ear vein and the microdialysis probe placed in the jugular vein. See text for additional details.

complexity of the blood dialysate. The negative ion MS/MS spectrum obtained from the collisonally activated decomposition of the (M-H)⁻ ion of penicillin G at *m/z* 333 shows an intense daughter ion at *m/z* 192. This fragment ion was used to monitor the concentration of drug in these experiments. Figure 4.21 compares the negative ion MS/MS spectra of penicillin G as a standard solution (c) and that obtained from the *in vivo* analysis of penicillin G in the blood of the rabbit (a) and the rat (b). The spectra are remarkably similar, showing a small amount of surviving parent at *m/z* 333 and the major fragment at *m/z* 192. The data for the on-line analysis of the drug in the rabbit are shown in Figure 4.22. Figure 4.22(a) shows the total ion chromatogram, where individual peaks represent individual injections of the perfusate into the mass spectrometer and Figure 4.22(b) shows the selected ion profile for *m/z* 192. One can see the expected positive hyperbolic infusion profile and negative hyperbolic elimination profile of the drug over a period of approximately 200

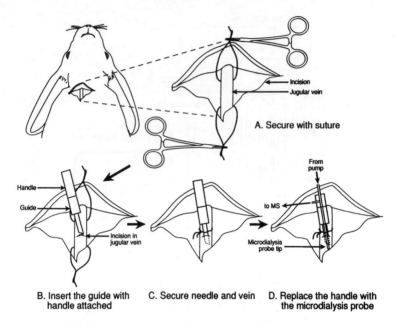

Figure 4.20. Surgical procedure for implanting the microdialysis probe into the jugular vein of the rabbit.

minutes. Figure 4.23 shows the results of two experiments, one for the intravenous administration of the drug in the rabbit and the other for the intramuscular injection of the drug in the rat (16). It was concluded from these initial experiments that the combination of microdialysis and mass spectrometry offers a very effective means of performing pharmacokinetic analysis of drugs and other endogenous metabolites in a single live animal.

4.4.3 Enzyme Reaction Monitoring

The use of microdialysis/MS for following concentrations of compounds in enzymic reactions was also investigated (15). Two methods of monitoring the action of carboxypeptidase Y on the substrate bombesin were employed. Figure 4.24(a) shows the results for the continuous analysis of the perfusate and Figure 4.24(b) that for perfusate that was first passed through a C-18 column to preconcentrate the analytes and then eluted with a one-step acetonitrile gradient. The reaction solution contained 250 pmol/μl of bombesin in 100 μl of solution. Although the appearance of the truncated molecular species of the polypeptide can be seen in the experiment utilizing direct

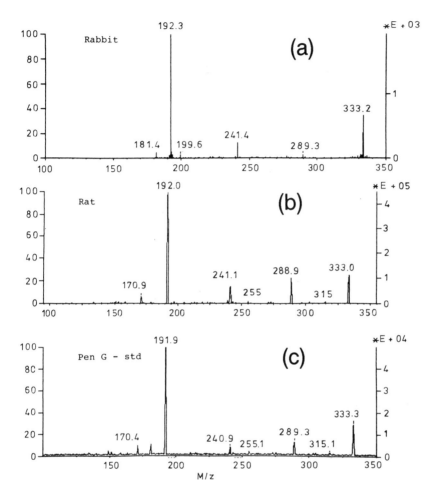

Figure 4.21. MS/MS daughter ion spectra of penicillin G obtained with CF-FAB for (a) *in vivo* rabbit study, (b) *in vivo* rat study, and (c) standard aqueous drug solution.

analysis of the perfusate, the sensitivity is poor at this concentration of substrate because the relative recovery of the dialysis process is low for this peptide. The results using a preconcentration step are considerably better and this procedure is recommended in such experiments. The disadvantage of the latter method is that the time course information for the appearance of individual peptides is lost since they are all eluted from the column at one time. However, this is not a serious drawback since the sequence information is determined by the mass difference of the fragments and not their rate of appearance during the reaction.

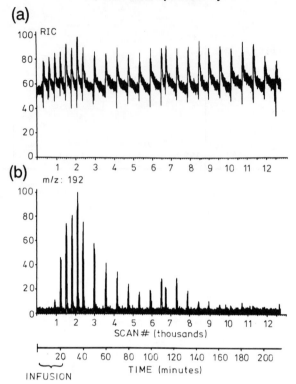

Figure 4.22. CF-FAB time-course analysis of dialysate in the *in vivo* rabbit study of penicillin G pharmacokinetics. (a) Total ion chromatogram, and (b) selected ion chromatogram of m/z 192, the most intense daughter ion in the MS/MS spectrum.

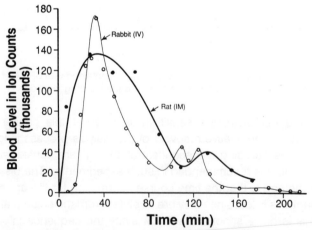

Figure 4.23. Pharmacokinetic analysis with microdialysis/CF-FAB for penicillin G administered either intravenously to the rabbit or intramuscularly to the rat.

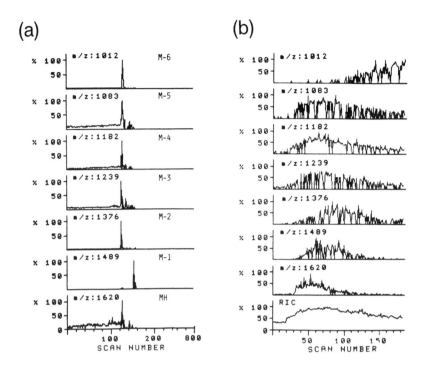

Figure 4.24. On-line reaction monitoring with microdialysis/MS of the action of carboxypeptidase Y on bombesin (MW 1619 daltons). (a) Desorption of reaction products from a trapping column (C-18 microbore column) using acetonitrile, and (b) constant flow of the perfusate at 4 µl/min.

4.5 CONCLUSION

The ability to directly monitor specific compounds in on-going reactions taking place in aqueous solutions is an extremely important asset to the biologist and chemist. Although several mass spectrometric techniques can be used to directly analyze aqueous solutions, CF-FAB has been shown to be extremely useful in a wide variety of applications. The mass specific nature of the detection accomplished by mass spectrometry provides the investigator with new capabilities with which to study biological systems. The implications to health and medicine are significant. Mass spectrometry is already an important analytical tool in helping unravel the mysteries of the living cell and new advances promise to accelerate this use. Moreover, it is believed that mass spectrometry will eventually contribute to the current trend in which vital tests on patients are performed at their bedside.

REFERENCES

1. R.M. Caprioli, T. Fan, and J.S. Cottrell, *Anal. Chem. 58*, 2949 (1986).
2. R.M. Caprioli, *Mass Spectrom. Rev. 6*, 237 (1987).
3. R.M. Caprioli, and T. Fan, *Biochem. Biophys. Res. Commun. 141*, 1058 (1986).
4. R.M. Caprioli, *Biochemistry 27*, 513 (1988).
5. W.E. Seifert, A. Ballatore, and R.M. Caprioli, *Rapid Commun. Mass Spectrom. 3*, 117 (1989).
6. R.M. Caprioli, *Biomed. and Environ. Mass Spectrometry 16*, 35 (1988).
7. R.M. Caprioli, and L. Smith, *Anal. Chem. 58*, 1080 (1986).
8. B. Whaley, and R.M. Caprioli, *Proceedings of the 37th ASMS Conference on Mass Spectrometry and Allied Topics*, Miami Beach, FL, 21-28 May 1989.
9. T. Zetterstrom, T. Sharp, C.A. Marsden, and U. Ungerstedt, *J. Neurochemistry 41*, 1769 (1983).
10. B.H.C. Westerling, G. Damsma, H. Rollema, J.B. DeVries, and A.S. Horn, *Life Sciences 41*, 1763 (1987).
11. S.A. Wages, W.H. Church, and J.B. Justice, Jr., *Anal. Chem. 58*, 1649 (1986).
12. S.N. Lin, and R.M. Caprioli, *Proceedings of the 36th ASMS Conference on Mass Spectrometry and Allied Topics*, San Francisco, CA, 5-10 June 1988.
13. A.K. Collin, and U. Ungerstedt, in *Microdialysis, User's Guide*, 4th edition (1988).
14. T. Huang, and P.T. Kissinger, *Current Separations 9*, 9 (1988).
15. S.N. Lin, S. Chang, and R.M. Caprioli, *Proceedings of the 38th ASMS Conference on Mass Spectrometry and Allied Topics*, Tucson, AZ, 3-8 June 1990.
16. R.M. Caprioli, and S.N. Lin, *Proc. Natl. Acad. Sci. 87*, 240 (1990).
17. R.M. Caprioli, *Anal. Chem. 62*, 477A (1990).
18. R.M. Caprioli, *Trends Anal. Chem. 7*, 328 (1988).
19. R.M. Caprioli, in *Biologically Active Molecules*, edited by U. Schlunegger, p. 59, Springer-Verlag, Berlin, 1989.

Chapter 5

LIQUID CHROMATOGRAPHY/MASS SPECTROMETRY

Richard M. Caprioli and Kenneth Tomer

Liquid chromatography has proven to be a very effective technique for the separation of an extremely diverse array of compounds. When coupled with mass specific detection by mass spectrometry, the instrument becomes a multi-dimensional analyzer for the identification of the masses of specific compounds as they elute from the chromatograph. Moreover, even under poor chromatographic resolution where mixtures of compounds elute together, individual compounds are easily identified as long as their molecular weights are different. Considerable effort has been made to develop an efficient and reliable interface to couple mass spectrometry with liquid chromatography. Many techniques have been successfully tested and applied in solving analytical problems, including thermospray, direct liquid introduction, electrospray, continuous-flow FAB, frit-FAB, atmospheric pressure ionization, particle beam and moving belt transport, as well as others. This chapter will focus on the use of continuous-flow (CF)-FAB as an interface for LC/MS.

The first attempt to couple FAB with LC utilized a moving belt interface (1). The technique involved the deposition of the eluant from the LC column onto a closed-loop stainless steel or polymer belt and subsequent transport of this region of the belt through desolvation chambers and into the source of the mass spectrometer. The appropriate portion of the belt containing the semi-dry sample spot was bombarded by the fast atom beam and the resulting ions were then analyzed. Although several applications have been reported using the belt technique, the method generally suffers from poor sensitivity and is mechanically complex.

Ito and his colleagues reported the flow of the liquid effluent from an LC directly into a FAB source through a fused-silica capillary (2). The capillary terminated in the source with a fine mesh or frit to disburse the sample and the glycerol included in the eluting solvent. This technique, termed frit-FAB, was used to separate and analyze the bile salts sodium taurocholate and sodium taurodeoxycholate on an ODS-Hypersil-5 (0.26 mm i.d. x 50 mm) column.

The mobile phase consisted of glycerol/acetonitrile/1.0 mM ammonium bicarbonate solution (10:22.5:67.5) which was allowed to flow into the source at a rate of 0.5 µl/min. This technique has subsequently been used to analyze several different types of compounds (3,4), and more recently has been employed to analyze protein digests (5). The latter application will be discussed in more detail later.

Continuous-flow FAB has been used as an LC/MS interface and operates by transporting liquid effluent from an LC column directly into the ion source for bombardment without the use of any dispersion devices. Typically, applications involving reverse-phase chromatography utilize an aqueous solvent containing 5% glycerol and an aqueous/organic solvent such as acetonitrile in either isocratic or gradient applications. The interface is able to handle flow rates between 1 and 20 µl/min, although optimum flow rates are in the 5 to 10 µl/min range.

The use of CF-FAB as an LC interface offers several important capabilities to the analyst. When used with microbore or capillary bore techniques, the method is very sensitive because compounds are eluted at high concentrations as a result of the low flow rates of solvent passing through the column. For example, the separation and identification of peptides in a tryptic digest can be done routinely at the 1-20 pmol level. Coupled with the low background characteristic of CF-FAB, mass spectra are produced with excellent signal-to-noise ratio measurements (6,7). The technique has a practical upper mass limit characteristic of that of FAB process, i.e., approximately 6,000 daltons using 8 keV bombarding atoms. It is compatible with high-voltage sources (8 - 10 keV) and can be used with a wide variety of solvents, including both aqueous and organic liquids. The interface is easily installed and removed in the instrument using the standard direct probe insertion lock of the mass spectrometer.

CF-FAB is best coupled with microbore and capillary bore columns (8,9). One of the basic parameters, which affects sensitivity, is the rate of liquid flow the interface is able to handle, and this specifies the split ratio of the column effluent for a given column size. Although full-bore columns have been used at elution rates of 500 - 1000 µl/min, split ratios of 1/50-1/100 had to be used so that only approximately 5-10 µl/min flowed into the ion source of the mass spectrometer (10). This high split ratio effectively lowers the overall sensitivity and limits full-bore applications to cases where sensitivity is not an issue or where separated compounds are to be collected. Microbore columns can be used at the 20 - 50 µl/min flow rate range and therefore would require a split ratio of not more than 1/3-1/4. Capillary bore packed columns (0.1 - 0.3 mm i.d.) are designed to run at flow rates of approximately 0.5-2 µl/min and requires no effluent splitting for the MS interface. Advantages and limitations of nanoscale (capillary) columns are discussed later in this chapter.

5.1 MICROBORE/MS APPLICATIONS

One of the most common applications of LC/MS is for peptide mapping, a technique which consists of the chromatographic separation and mass spectrometric detection of a mixture of peptides produced by the action of a protease on a substrate protein. Reaction aliquots can be directly injected onto a reverse-phase microbore column, commonly 1 x 50 mm or 1 x 250 mm, without deproteinization or preconcentration (11-12). The column is eluted with an acetonitrile gradient over a period of 15-30 minutes. Any sacrifice of column resolution because of the lower flow rate used or rapid gradient formation is not serious because the mass-specific nature of the detector allows one to trace individual masses independent of other masses in the spectrum.

A typical microbore LC/MS set-up for CF-FAB is shown in Figure 5.1. A commercially available microbore LC may be used (depicted inside the dotted line in the figure) and generally includes a UV detector with a micro (<1 µl) cuvette. A splitter has been added to the system between the syringe pump and the injector so that gradients may be formed at relatively high rates (500

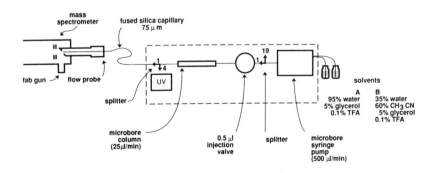

Figure 5.1. Instrumental set-up for microbore LC/MS with the CF-FAB interface. (Reprinted with permission from reference 9.)

μl/min or more). A 1/19 splitter then allows approximately 25 μl to flow through the microbore column. A second splitter, approximately 1/4, is used after the column to transmit 5 μl/min to the CF-FAB interface and 20 μl/min to the UV detector. For peptide separations using a C-18 reverse-phase microbore column, 5% (V/V) of glycerol is added to both solvents. Generally, solvent A consists of 95% water, 5% glycerol, 0.1% TFA, and solvent B, 35% water, 60% acetonitrile, 5% glycerol, and 0.1% TFA. In applications where the addition of glycerol or other FAB matrix to the solvent may compromise column performance, this matrix may be added post-column as described later in this chapter.

The slope of the elution gradient provides a convenient means of controlling the total analysis time. Although rapid separations are desirable, they also usually lead to poor column resolution. Figure 5.2 illustrates the efficiency of the chromatographic separation for three column sizes, 2 x 2 mm, 2 x 10 mm and 1 x 50 mm RP-300 C-8 (Applied Biosystems), achieved when the first segment of the gradient profile is lengthened from 2.4 minutes to 9.9 minutes

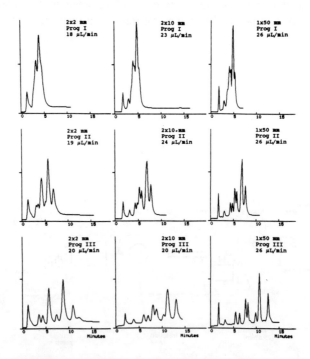

Figure 5.2. Resolution of peptides from the tryptic digest of cytochrome c as a function of column size (ABI RP-300 C-8) and gradient conditions. The vertical panels represent column dimension of (left) 2 x 2 mm, (middle) 2 x 10 mm, and (right) 1 x 50 mm. Gradient programs all were 5 - 40% B in 2.4 min, 4.9 min, and 9.9 min for programs I, II, and III, respectively. (Reprinted with permission from reference 9.)

at flow rates of about 20 µl/min (9). The results show that a wide variation of column performance can be obtained and that this can be used and tailored to the particular need of the analyst. Of course, the trade-off here is analysis time versus column resolution. The mass spectrometer, however, allows "recovery" of much of the resolution lost with short analysis times through the use of selected ion chromatograms where individual masses can be monitored. Thus, if the resolution of the sample is basically to be obtained from the mass spectrum and only separation from the solvent front containing buffer and salt is desired, then a small cartridge (e.g., 2 x 2 mm) may give satisfactory performance in the time/resolution trade-off. On the other hand, if greater column resolution is needed because, for example, fractions need to be collected for further studies, then a 1 x 50 mm or 1 x 250 mm column would undoubtedly be preferred. Overall, the choice of the chromatographic conditions used in the analysis of a particular sample depends upon the complexity of the sample and the data desired by the analyst.

The tryptic digest of human apolipoprotein A1 is an example of a complex peptide mixture analyzed by CF-FAB LC/MS which utilizes a medium-length column (13). Figure 5.3 shows the total ion chromatogram for the analysis of

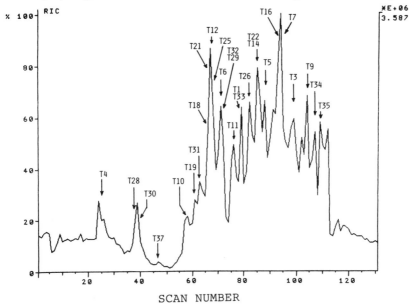

Figure 5.3. Total ion chromatogram for the LC/CF-FABMS analysis of 24 pmol of the tryptic digest of human apolipoprotein A1. The tryptic fragments identified from the mass spectra are indicated.

24 pmol of the protein performed on a 1 x 50 mm RP-300 C-8 column eluted with a 0-50% acetonitrile gradient over a period of about 30 minutes. Although only the major tryptic fragments are labeled in this figure, over 80 individual peptide molecular species were identified, most of which were derived from low-level protein contaminants in the sample. Figure 5.4 shows the detailed

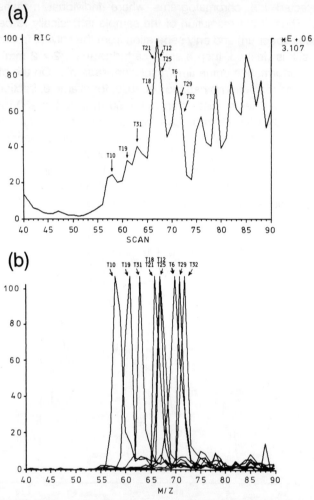

Figure 5.4. (a) Portion of the total ion chromatogram shown in Figure 5.3. The tryptic peptides between scans 55 and 75 are noted. (b) Selected ion chromatograms of the $(M+H)^+$ ions for peptides present in scans 55 - 75. (Reprinted with permission from reference 27.)

mass spectral analysis of a portion of this chromatogram between scans 56 and 73, and the selected ion chromatograms of each of the peptide molecular species identified in this region. The latter chromatogram was produced by plotting each selected ion chromatogram, independently normalized, on the same axis. This demonstrates quite clearly the advantage of mass specific detection in allowing the analyst to isolate specific species even though they may be present in complex chromatographic peaks.

The mass spectra obtained using CF-FAB are characterized by intense $(M+H)^+$ ions and low background, especially above m/z 400. Figure 5.5

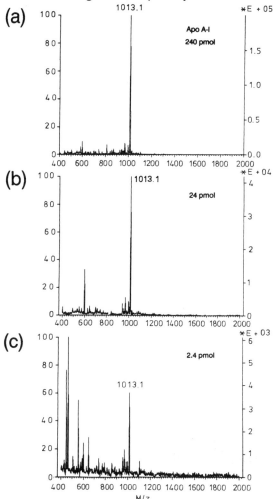

Figure 5.5. Mass spectra of the tryptic fragment of m/z 1013 from the analysis of 2.4, 24, and 240 pmol of the digest of apolipoprotein A1. (Reprinted with permission from reference 15.)

shows the mass spectra of the tryptic fragment at m/z 1013 obtained from a series of analyses representing the hydrolysis of 240, 24, and 2.4 pmoles of apolipoprotein A1. Even at the 2.4 pmol level, the signal-to-noise ratio for $(M+H)^+$ is excellent. It is noted that the sample amounts shown in the figure are overestimates. Actual injected amounts are probably significantly less due to contaminants in the original protein and the lack of complete hydrolysis of the protein by trypsin.

A "mass summary", i.e., a mass spectrum produced by summing all of the individual spectra in the analysis, is shown in Figure 5.6(b). This is a

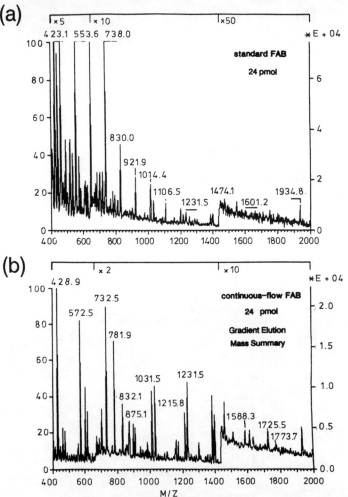

Figure 5.6. Comparison of mass spectra from the tryptic digest mixture of human apolipoprotein A1 with (a) standard FAB, and (b) LC/CF-FAB with summed mass spectra over the chromatogram range shown in Figure 5.3.

convenient way to observe the total fragment array and catalogue specific masses identified, although this presentation does not accurately reflect the signal-to-noise ratio of individual measurements. It is interesting to compare this mass summary spectrum to the mass spectrum obtained with standard FAB analysis (normally used for peptide mapping) of the same digest mixture of apolipoprotein A1, shown in Figure 5.6(a). In the standard FAB spectrum, the most intense ions are derived from glycerol cluster ions and the peptide molecular species are present at relatively low intensities. In contrast, the mass summary spectrum obtained with LC/CF-FAB shows intense peptide molecular species with background and glycerol cluster ions of low intensities. Table 5.1 lists the peptide fragments expected from this proteolytic digest, the

Table 5.1. Apolipoprotein A-1 Tryptic Fragments

	Sequence	$(M+H)^+$ calc.	$(M+H)^+$ meas.
1	DEPPQSPWDR	1226.5	1226.9
2	DLATVYVDVLK	1235.7	1236.2
3	DSGR	434.2	434.2
4	DYVSQFEGSALGK	1400.6	1400.9
5	QLNLK	615.4	615.4
6	LLDNWDSVTSTFSK	1612.8	1613.8
7	EQLGPVTQEFWDNNLEK	1932.9	1933.9
8	ETEGLR	704.3	704.7
9	QEMSK	622.3	622.6
10	DLEEVK	732.4	732.7
11	VQPYLDDFQK	1252.6	1252.9
12	WQEEMELYR	1283.6	1284.0
13	VEPLR	613.3	613.3
14	AELQEGAR	873.4	873.9
15	LHELQEK	896.5	897.0
16	LSPLGEEMR	1031.5	1031.5
17	AHVDALR	781.4	781.8
18	THLAPYSDELR	1301.6	1301.9
19	LAAR	430.3	430.1
20	LEALK	573.3	573.6
21	ENGGAR	603.3	603.5
22	LAEYHAK	831.4	831.8
23	ATEHLSTLSEK	1215.6	1215.8
24	AKPALEDLR	1012.6	1013.1
25	QGLLPVLESFK	1230.7	1231.1
26	VSFLSALEEYTK	1386.7	1387.1
27	LNTQ	475.2	475.2

calculated $(M+H)^+$ values, and the measured $(M+H)^+$ values obtained by LC/CF-FAB. In the mass range scanned (m/z 400-2000), all of the peptides expected were identified. It is clear that the LC/MS technique is the method of choice for the identification of peptides in protease digests.

Peptide mapping also provides a means to detect low-level protein contaminants for protein samples from biological origin or those prepared synthetically. For example, Figure 5.7 shows the sequence of recombinant

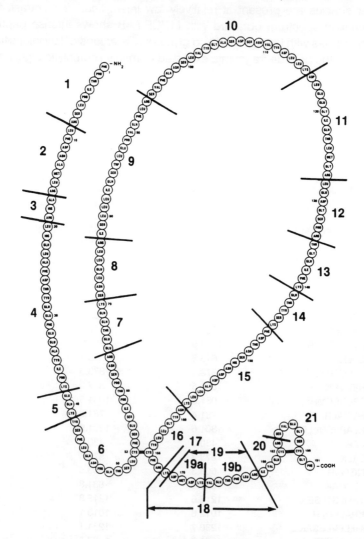

Figure 5.7. Sequence of human growth hormone showing the sites of the tryptic cleavages. (Reprinted with permission from reference 14.)

human growth hormone and the positions of the various tryptic fragments (14). To study the sensitivity of LC/MS in identifying contaminants at low levels, a sample of this protein was obtained which was known to contain approximately 1% of the protein with a single methionyl residue oxidized. The total ion chromatogram for the analysis of the tryptic digest is shown in Figure 5.8 (15).

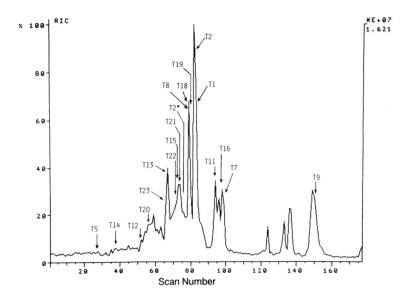

Figure 5.8. Total ion chromatogram of the microbore LC/CF-FABMS analysis of the tryptic digest of recombinant human growth hormone. The tryptic fragments identified from their mass spectra are indicated, including T2*, the oxidized T2 peptide.

Two tryptic fragments should be noted; T2 and T2*. The latter is seen as a small shoulder on another peak in the chromatogram and was identified from the mass spectral analysis of these data. Figure 5.9 shows the molecular ion region of the peptides found in the chromatographic peaks; (a), the expected molecular species for T2 at m/z 980 and (b), the oxygenated analog T2* at m/z 996. Comparison of the intensities shown on the right-hand side of the spectrum, assuming equal ionization efficiency, shows that there is approximately 1.4% of the monosulfoxide analog present in the original sample. From these data, it is concluded that peptide contaminants in protease digests can be observed at the 0.5% level in an analysis where approximately 200 pmol of the original protein has been analyzed.

Carr and coworkers (16) have studied the structure of recombinant soluble CD4 receptor. This receptor, a 55 kDa glycoprotein, has been shown to directly interact with class II major histocompatibility complex antigens,

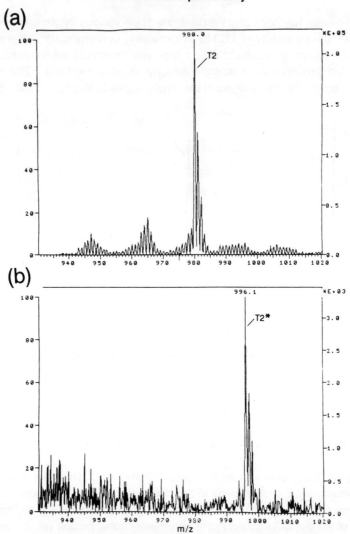

Figure 5.9. Mass spectra from the analysis of recombinant human growth hormone (see Figure 5.8) showing the $(M+H)^+$ region in the mass spectrum of (a) T2 (fragment 9 - 16) and (b) T2*, the oxidized form. (Reprinted with permission from reference 15.)

mediating the immune response. CD4 also serves as the receptor for the human immunodeficiency virus (HIV). These workers used mass spectrometry and corroborated 95% of the 369 amino acid long sequence. LC/MS was used to analyze the tryptic digest of this protein using microbore techniques and CF-FAB. An aquapore RP-300 column (1 mm x 100 mm C-18) was used with gradient elution, where solvent A is 5% glycerol and 0.1% TFA in water,

Liquid Chromatography/Mass Spectrometry

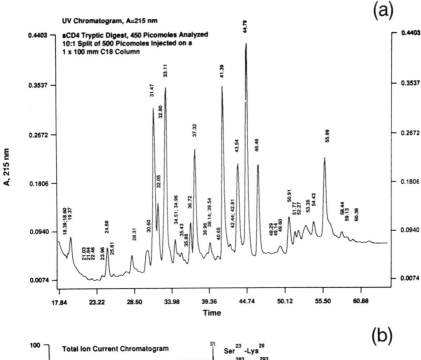

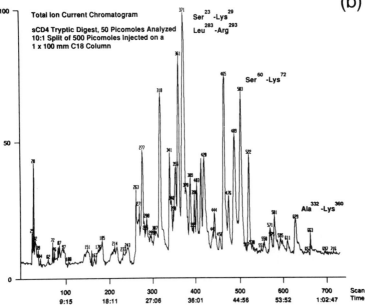

Figure 5.10. Analysis of the tryptic digest of recombinant soluble CD4 receptor protein; (a) UV detector trace; (b) total ion chromatogram, using on-line microbore LC/CF-FAB. (Reprinted with permission from reference 16.)

and solvent B is 60% acetonitrile, 5% glycerol and 0.1% TFA in water. The gradient was set up for T=0 → 5 min, 7% B; T=5-16 min, 7-19% B; T=16-50 min, 19-60% B, and T=50-100 min, 60-100% B. A sample of a 6 hour tryptic digest of the protein containing 500 pmol in 8 µl was injected onto the column, which was eluted at a rate of 40 µl/min. A 10:1 split ratio was used, providing a flow of approximately 3.3 µl/min into the mass spectrometer. A VG ZAB SE-4F tandem MS was used with a FAB ion gun and was scanned from m/z 4050-850 at 7 s/decade and was set at a resolution of 500. Figure 5.10 shows both the UV and MS analysis of this sample. For simplicity, only four of the peptides identified are indicated in this figure, with their mass spectra shown in Figure 5.11. The signal-to-noise ratio is good in these spectra, despite the relative small sample split into the MS (about 50 pmol) and the extremely wide mass range scanned.

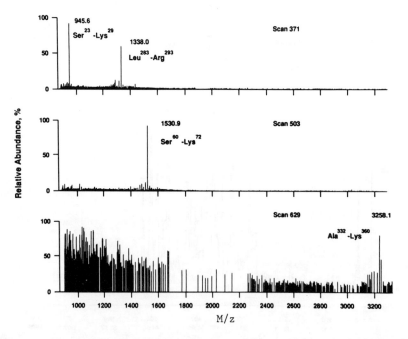

Figure 5.11. Mass spectra of several peptides identified in the total ion chromatogram shown in Figure 6.10. (Reprinted with permission from reference 16.)

Bell et al. (5) also used tandem MS with frit-FAB and a microbore RP-300 column under conditions similar to that used above to analyze the tryptic digest of Lys-78-plasminogen, a protein of molecular weight of about 84 kDa which produces 70 unique tryptic fragments. Approximately 40 pmol of the digested protein was injected and the column eluted at 200 μl/min. A solution of 2% glycerol in methanol was added post-column at 200 μl/min and the entire flow split to allow 13 μl/min to flow into a JEOL SX-102 MS equipped with a FAB gun. The scan cycle time was 10 seconds over a mass range of 50-2400. The averaged and background subtracted MS/MS spectrum for a peptide at m/z 1238 is shown in Figure 5.12. The fragmentation observed is

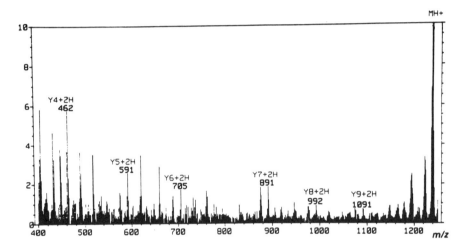

Figure 5.12. MS/MS (collisionally activated decomposition) mass spectrum from the $(M+H)^+$ ion of one peptide identified in the tryptic digest of the protein Lys-78-plasminogen using microbore LC/frit-FAB techniques. (Reprinted with permission from reference 4.)

predominantly of the (Y+2H) series which is directed by the C-terminal arginine of this peptide. The background subtraction capability for such on-line analyses is an important advantage.

5.2 CAPILLARY (NANOSCALE) LC/MS

5.2.1 The Co-axial Interface

The use of LC columns with inner diameters less than 400 μm is becoming more prevalent due to the smaller sample load, decreased consumption of solvent and generation of solvent waste, and the increased separation efficiencies associated with these columns. The traditional incorporation of the FAB matrix, commonly glycerol, into the mobile phase can create problems with columns having small inner diameters. To maintain the same amount of glycerol at the probe tip, the concentration of glycerol in the mobile phase must be increased due to the low flow rates used with these columns and this can seriously affect chromatographic performance. For very narrow-bore columns, such as 50 μm (i.d.) packed columns or 10 μm (i.d.) open tubes used with flow rates of about 50 nl/min, incorporation of glycerol in the mobile phase provides insufficient glycerol to the probe tip. Post-column addition of the matrix can also cause problems and leads to unacceptable band-broadening in some cases.

Tomer and coworkers (17, 19) have developed a coaxial interface for separately delivering the matrix and the analyte to the probe tip in which the analytical column is inserted inside a larger-diameter column which delivers the FAB matrix. This system consists of a syringe pump to deliver the matrix, an injection system, a tee for connecting the matrix flow, an analytical column, a sheath column, and a CF-FAB probe interface. Typical operating conditions for either open tubular columns or packed nanoscale capillary columns are as follows:

Open Tubular Column; 10 μm i.d./150 μm o.d.
Packed Microcapillary Column; 50 μm i.d./150 μm o.d.
Microcapillary Column Packing; 10 μm C18
Column Flow Rates; 50 nl/min
Sample Injection Volumes; 0.5 to 10 nl
Sheath Column; 160 μm i.d./350 μm o.d.
FAB Matrix Composition; 25/75 Glycerol/Water
FAB Matrix Flow Rate; 1 μl/min

The FAB matrix solution is delivered to the system using a 1/16 inch stainless steel Swagelock tee, as shown in Figure 5.13. The analytical column is pushed through the tee into the sheath column towards the mass spectrometer, terminating several millimeters from the end of the sheath column. The FAB matrix solution is supplied by a pump through a capillary that is connected to the side arm of the tee. Vespel ferrules (0.4 mm i.d.) are used to hold the columns in the tee. Graphite ferrules are not recommended because they tend to shed and plug the column. Insertion of the analytical column into the sheath is facilitated by first wetting the sheath column with water.

Several CF-FAB probe interfaces of different designs have been successfully used. One version, constructed in our laboratory, is illustrated in Figure 5.14 and uses a sheath column that is flush with the FAB probe tip.

Liquid Chromatography/Mass Spectrometry

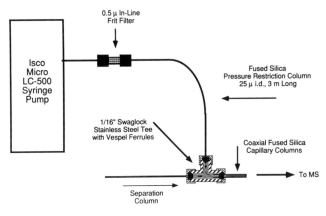

Figure 5.13. Schematic diagram of the FAB matrix delivery system and the interface tee for the coaxial CF-FAB interface.

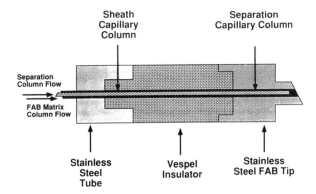

Figure 5.14. Coaxial continuous-flow FAB probe tip. (Reprinted with permission from reference 17.)

The analytical column is withdrawn 1-2 mm into the sheath column in order to prevent the vacuum in the mass spectrometer from evaporating mobile phase within the column. The fit between the outer diameter of the sheath column and the inner diameter of the hole in the probe tip should be as tight as possible to prevent back-flow of the matrix. Placement of a septum, Torr Seal or PTFE tape (20) in the probe tip to prevent back-flow is also recommended. The reader is referred to Chapter 1 for additional details.

The matrix liquid is delivered by a syringe pump (e.g., Isco Micro LC-500) capable of sustaining constant flow rates as low as 100 nl/min. A 0.5 μm in-line frit filter and a 3 m x 25 μm (i.d.) fused silica capillary acts as a pressure-restriction column, providing sufficient back-pressure for stable operation of the syringe pump. Operating conditions listed above are typical, but the maintenance of a stable matrix liquid surface on the target is dependent on the temperature of the probe tip, surface area of the probe tip, and composition of the matrix flow. To initially determine optimum flow rates, a series of injections of a standard should be made, for example, the tripeptide Met-Leu-Phe. The optimum flow will provide greatest sensitivity and a narrow peak shape. Excessive flows will lead to a broadened peak shape due to diffusion of the analyte into the matrix drop and too little flow will provide insufficient matrix to wet the probe tip leading to decreased sensitivity. Once an optimum flow is determined, the source vacuum gauge reading provides a simple device to monitor and maintain stable conditions. As in all LC experiments, all liquids should be filtered and exhaustively degassed, and no air pockets should remain in the system.

5.2.2 Sample Injection

Three injection systems have been developed for nanoscale capillary columns: splitless/split, pressurized injection, and microinjection.

The simplest method is splitless/split injection depicted in Figure 5.15. This system consists of a standard Valco six-port valve connected to the mobile phase reservoir with 1/16 inch stainless steel tubing and to a 1/16 inch stainless steel tee with 1/16 inch stainless steel tubing and an in-line filter. The nanoscale capillary column fits into the other arm of the tee while the side-arm of the tee is connected to a sample recovery vial through an on/off valve. To make a sample injection, the volume between the Valco valve and the sample recovery valve (including the splitter tee) is filled with analyte solution. The recovery valve is closed and the Valco valve is switched to permit flow of the mobile phase into the system. After a predetermined injection time, typically 5 seconds at a column flow rate of 1 nl/s, the recovery valve is opened and the stainless steel tubing and splitter tee are rinsed out with mobile phase. After

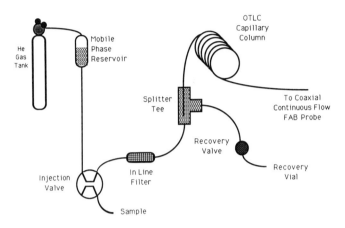

Figure 5.15. Splitless/split injection system. (Reprinted with permission from reference 18.)

the system is rinsed out (approximately 2-3 seconds) the recovery valve is closed and chromatography commences. The total volume of analyte solution needed is approximately 250 μl. A smaller volume (about 50 μl) can be used if the stainless steel lines between the injection valve and the splitter tee are replaced with fused silica (250 μm i.d.). It is noted that only about 5 nl of the 250 μl injection solution is consumed and the remainder is recovered.

If the volume of analyte solution is greater than about 10 μl, a pressurized injection system can also be used (Figure 5.16). The pressurizable stainless steel vessel is built to hold a 1 ml conical vial. The end of the nanoscale capillary is placed through the fitting on the top of the vessel and inserted into the analyte solution. The top of the vessel and the column fitting are tightened and the vessel pressurized by helium at the same pressure used to pressurize the mobile phase. After the system has been pressurized for the predetermined time (the same as for the splitless/split injection), the pressure is released. The column is removed from the vessel and replaced into the splitter tee and chromatography begun by permitting flow of the mobile phase.

Picoliter injection volumes can be made using a microinjector (21), although this system is generally too laborious for common use. Briefly, the technique involves loading the tip of a very finely drawn capillary with sample solution and, using a microscope, placing this capillary tip into the end of the capillary chromatography column. The sample plug is then pushed onto the head of the column with low pressure.

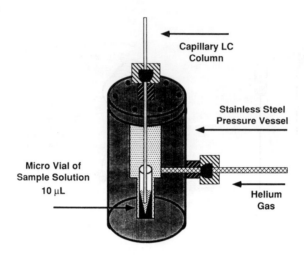

Figure 5.16. Pressurized injection vessel for capillary LC.

Correlation of the chromatograms obtained from the splitless/split and the microinjector systems is quite good. Figure 5.17 shows the selected ion trace of the $(M+H)^+$ ion of dopamine (0.4 pmol) injected by the two methods. The total ion counts for both traces is well within the experimental error of the techniques (1.89×10^6 counts for the splitless/split and 1.96×10^6 counts for the microinjector). The total volume injected was 14 nl for the splitless/split injection and 4 nl for the microinjector, with each sample volume containing 0.4 pmol of dopamine.

5.2.3 Chromatographic Parameters

The simplest system for delivering the eluting solvent isocratically is the same as that shown earlier in Figure 5.15. The mobile phase is placed into a reservoir that is pressurized by helium. Gradient solvent delivery is accomplished using the system shown in Figure 5.18, designed primarily for use with packed 50 μm (i.d.) columns. A conventional HPLC pumping system is used in place of the mobile phase reservoir. A flow restrictor and flow splitter/restrictor are needed to split the 1 ml/min flow rate from the pumps down to the nl/s flow rate required by the capillary system. The first restrictor is used to maintain back-pressure on the pumping system during injection of the sample, and the second splitter/restrictor is used to split off the excess

Liquid Chromatography/Mass Spectrometry

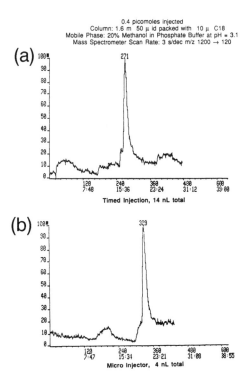

Figure 5.17. Selected ion traces of the $(M+H)^+$ ion of 0.4 μmol of dopamine injected by (a) a timed injection and (b) by the microinjector.

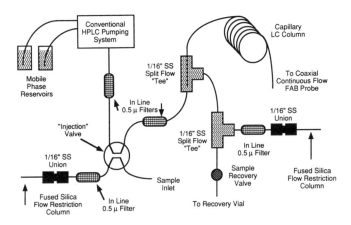

Figure 5.18. Gradient solvent delivery system for 50 μm (i.d.) packed capillary columns.

mobile phase flow. The length of restrictors needed can be calculated from Poiseuille's Law:

$$F = (\pi r^4 P)/8\eta L$$

where F is the flow rate, r is the column radius, P is the pressure differential, η is the fluid viscosity, and L is the column length. In the present set-up, a 3 cm x 25 μm (i.d.) restrictor column gives satisfactory performance.

Sensitivity of the coaxial arrangement is very good, presumably because the mobile phase flow is concentrated at the probe tip and is somewhat constrained from dispersion across the whole target surface by the surrounding matrix flow. This apparently leads to increased concentration at the center of the probe with concomitant increased sensitivity. Full-scan spectra for a variety of peptides have been obtained on less than 1 ng of sample (and as low as 22 pg) and on low nanogram amounts of phospholipids, saccharides, and steroids. For example, Figure 5.19 shows the

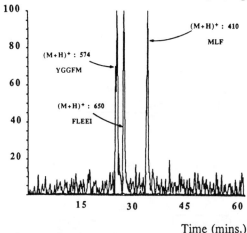

Figure 5.19. Selected ion chromatograms from the LC/CF-FAB MS analysis of 2 picomoles each of Tyr-Gly-Gly-Phe-Met, Phe-Leu-Glu-Glu-Ile, and 3 picomoles of Met-Leu-Phe. A 50 μm (i.d.) x 2.2 m capillary column packed with 10 μm C-8 particles was used with a gradient elution of acetonitrile/water/TFA (30/70/0.1) → (90/20/0.1) in 15 minutes at a flow rate of 86 nl/min. The matrix solution was 25% glycerol in water delivered at a flow rate of 1 μl/min.

separation of a mixture of three peptides on a 50 μm (i.d.) x 2.2m column packed with 10 μm (C-8) particles. The gradient was acetonitrile:water:TFA (30:70:01) to (90:20:01) in 15 minutes at a flow rate of 86 nl/min. Approximately 2 pmol of each peptide were loaded and a matrix solution containing glycerol:water (25:75) was used at a flow rate of 1 μl/min. Sub-picomolar amounts of tryptic digests have also been successfully analyzed (22). In addition, the coaxial arrangement provides a simple and routine method for introduction of analytes for acquisition of MS/MS data (23).

5.2.4 Other Applications

A number of other workers have also used capillary CF-FAB LC/MS in several different applications. Ackermann, et al. (24) have used MS/MS for the LC/MS analysis of peptide digests to obtain sequence information from the individual peptides eluting from the column. These workers used a 0.320 mm (i.d.) x 25 cm fused silica column slurry packed with 5 μm Spherisorb ODS particles. The column was eluted at a flow rate of 2 μl/min with a solution containing acetonitrile:water:glycerol:thioglycerol (4:2:1:1) added post-column at a rate of 1 μl/min. Daughter ion spectra were recorded using B/E linked scanning at a parent ion attenuation of 30% with helium as the collision gas. Figure 5.20(a) shows the elution profile of the tryptic fragments of glucagon from the analysis of 290 pmol of the reaction mixture. This separation was achieved using a linear gradient from 5-60% acetonitrile in water containing 0.1% TFA over a period of 30 minutes. Four tryptic fragments of glucagon were identified in this analysis. It should be noted that in the standard FAB analysis of this digest, only three peptides are identified, the fourth, at m/z 1354, is extremely hydrophilic and is not usually observed. Figure 5.20(b) shows the MS/MS analysis of fragment T3, a peptide with $(M+H)^+ = m/z$ 1352. Nearly complete sequence information was obtained from this spectrum from the 290 pmol injection of the peptide digest. Although the entire sequence of glucagon was not verified in this case, sequence ions were identified for most of the structure. It is clear that the technique can be used quite effectively in verifying the structure of proteins whose putative sequences have been inferred by other means or that have been partially sequenced by classical techniques.

Boulenguer, et al. (25) have used capillary bore CF-FAB LC/MS for the analysis of permethylated oligosaccharide alditols obtained from hen ovomucoid. The separation, shown in Figure 5.21, was preformed on a fused silica RoSil C-18 column (0.32 mm i.d. x 25 cm) at a flow rate of 5 μl/min. The elution was carried out with a gradient starting with 80% solvent A (methanol:thioglycerol:water 90:10:5) and 20% solvent B (thioglycerol:water 10:90) to 100% solvent A in 20 minutes. A mass range of m/z 3200-900 was scanned at a rate of 10 s/decade with a mass resolution of 3000. Several of the oligosaccharides molecular ions were identified in this mixture and are shown in the figure. This sample analysis represents approximately 4.5 μg of total permethylated oligosaccharide alditols injected onto the column. A new oligosaccharide was identified in this mixture and is indicated as the broken arrow in the figure at approximately scan 355. The molecular species was observed at m/z 3086 and was consistent with a hexaantennary oligosaccharide alditol consisting of galactose, mannose, N-acetylglucosamine, and N-acetylglucosaminatol in a molecular ratio of 1:3:8:1. Since no splitting

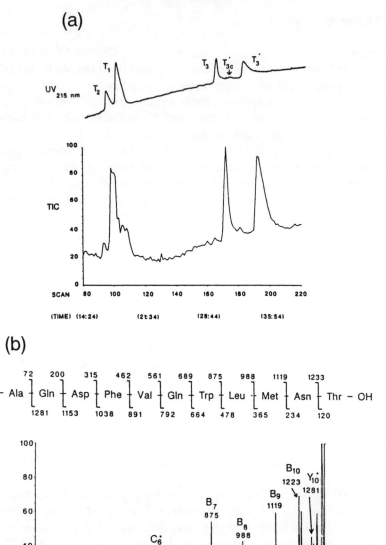

Figure 5.20. (a) The UV and total ion chromatogram from the LC/CF-FABMS analysis of 290 pmol of the tryptic digest of glucagon. (b) Daughter ion scan for the tryptic fragment T3, $(M+H)^+ = m/z$ 1352. Accepted peptide cleavage notation is used to denote the fragmentation pattern. (Reprinted with permission from reference 24.)

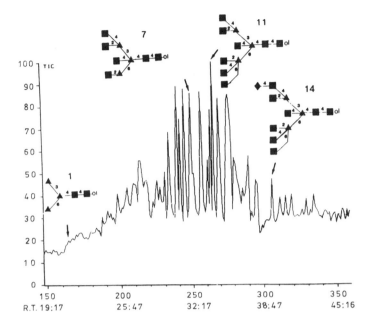

Figure 5.21. Total ion chromatogram from the LC/CF-FAB analysis of permethylated oligosaccharides obtained with a packed fused-silica column. (Reprinted with permission from reference 25.)

is required and all of the column effluent can be directed to the mass spectrometer, capillary bore LC/MS provides high sensitivity and good chromatographic performance. This has been demonstrated with a variety of samples, including herbicides and their plant metabolites and acylcarnitines in human urine. Additional details for these, as well as other applications, are given in Chapter 8.

Kokkonen and coworkers (26) have utilized both full-bore and microbore LC/MS for target compound analysis in analyzing dextromethorphan in human plasma. These workers have employed a technique called "phase-system switching" which allows them to use a full-bore column for the separation and a microbore column for concentration based on valve-switching techniques as shown in Figure 5.22. Basically, the compound of interest is trapped after eluting from the analytical column onto a short trapping column. After washing and drying of this trapping column, it is eluted with an appropriate solvent at a flow rate which allows the entire effluent to be transmitted to the CF-FAB interface. A glycerol solution was added post-column as the FAB matrix. An RP-2 analytical column (3 mm i.d. x 100 mm) was used at a flow rate of 1 ml/min in combination with an XAD-2 trapping column (1 mm i.d. x 50 mm).

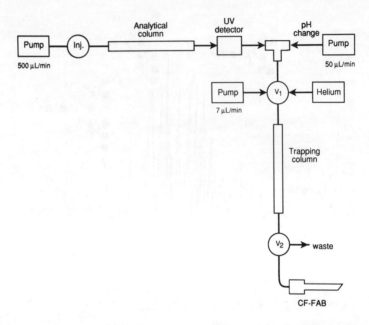

Figure 5.22. Schematic representation of the phase system switch technique for the analysis of target compounds. See text for details. (Reprinted with permission from reference 26.)

A pH change was made after the elution of the dextromethorphan from the analytical column in order to trap it on the XAD-2 column. The technique of phase-system-switching is an extremely effective analytical method for target compound analysis. It provides for the use of full-bore columns, and therefore injections of relatively large samples with the final determination by CF-FAB of the entire effluent from the trapping column which contains the sample of interest at high concentration without any splitting in the system.

5.3 CONCLUSION

LC/MS represents one of the most powerful analytical tools available today. The use of FAB ionization through the continuous-flow interface permits the analysis of a wide variety of charged and polar compounds without the need for derivatization or extensive sample preparation. The technique is ideally suited to microbore and capillary bore chromatography because of the low flow rates used. This also provides high sensitivity because compounds are eluted at high concentrations since elution volumes are small. The mass spectrometer acts as a multi-dimensional analyzer where each mass-to-charge value scanned is a separate specific data set, allowing individual compound identification in complex chromatographic peaks containing unresolved mixtures of compounds.

REFERENCES

1. J.G. Stroh, J.C. Gook, R.M. Milberg, L. Brayton, T. Kihara, Z. Huang and K.L. Rinehart, *Anal. Chem. 57*, 985 (1985).
2. Y. Ito, T. Takeuchi, D. Ishi and M. Goto, *J. Chromatogr. 346*, 161 (1985).
3. T. Takeuchi, S. Watanabe, N. Kondo, D. Ishii and M. Goto, *J. Chromatogr. 435*, 482 (1988).
4. Y. Ito, T. Takeuchi, D. Ishii, M. Goto and T. Mizuno, *J. Chromatogr. 391*, 296 (1987).
5. D.J. Bell, M.D. Brightwell, W.A. Neville, and A. West, *RCM 4*, 88 (1990).
6. R.M. Caprioli, T. Fan and J.S. Cottrell, *Anal. Chem. 58*, 2949 (1986).
7. R.M. Caprioli and T. Fan, *Biochem. Biophys. Res. Commun. 141*, 1058 (1986).
8. R.M. Caprioli, B. DaGue, T. Fan and W.T. Moore, *Biochem. Biophys. Res. Commun. 146*, 291 (1987).
9. R.M. Caprioli, B.B. DaGue and K. Wilson, *J. Chromatogr. Sci. 26*, 640 (1988).
10. D.E. Games, S. Pleasance, E.D. Ramsey and M.A. McDowell, *Biomed. Environ. Mass Spectrometry 15*, 179 (1988).
11. R.J. Simpson and E.C. Nice, *Methods in Protein Sequence Analysis 213* (1986).
12. R.M. Caprioli, W.T. Moore, B. DaGue and M. Martin, *J. Chromatogr. 433*, 355 (1989).
13. W.T. Moore, B. DaGue, M. Martin and R.M. Caprioli, *Proc. of the 36th Annual ASMS Conf. on Mass Spectrometry and Allied Topics*, 5-10 June, San Francisco, CA (1988).
14. G.W. Becker, P.M. Tackitt, W.W. Bromer, D.S. Lefeber and R.M. Riggin, *Biotech. and Applied Biochem. 10*, 326 (1988).
15. R.M. Caprioli, in *Biologically Active Molecules*, edited by U.P. Schlunegger, Springer-Verlag, Berlin, 79 (1989).
16. S. Carr, G.D. Roberts and W. Johnson, personal communication.
17. J.S.M. deWit, L.J. Deterding, M.A. Moseley, K.B. Tomer and J.W. Jorgenson, *Rapid Commun. Mass Spectrom. 2*, 100 (1988).
18. M.A. Moseley, L.J. Deterding, K.B. Tomer and J.W. Jorgenson, *J. Chromatogr. 480*, 197 (1989).
19. M.A. Moseley, L.J. Deterding, K.B. Tomer, R.T. Kennedy, N.L. Bragg and J.W. Jorgenson, *Anal. Chem. 61*, 1577 (1989).
20. S.J. Gaskell and R. S. Orkiszewski, Chapter 2, this volume.
21. R.T. Kennedy, R.L. St. Claire, III, J.G. White and J.W. Jorgenson, *Mikrochim. Acta 11*, 37 (1987).
22. S. Pleasance, P. Thibault, M.A. Moseley, L.J. Deterding, K.B. Tomer and J.W. Jorgenson, *J. Am. Soc. Mass Spectrom.* In Press.
23. L.J. Deterding, M.A. Moseley, K.B. Tomer, J.W. Jorgenson, *Anal. Chem. 61*, 2504 (1989).
24. B.L. Ackerman, T.M. Chen and J.E. Coutant, *Proc. 37th ASMS Conf. on Mass Spectrom. and Allied Topics*, 21-26 May, Miami Beach, FL 383 (1989).
25. P. Boulenguer, Y. Leroy, J. Alonso, J. Montreuil, G. Ricort, C. Colbert, D. Duquet, C. Dewaele and B. Fournet, *Anal. Biochem. 168*, 164 (1988).

26. P. Kokkonen, W.M. Niessen, U.R. Tjaden and J. van der Greef, *J. Chromatogr.* **474**, 59 (1989).
27. R.M. Caprioli, *Anal. Chem.* **62**, 477A (1990).

Chapter 6

CAPILLARY ZONE ELECTROPHORESIS / MS

Kenneth Tomer and M. Arthur Moseley

Electrophoretic separations are widely used in the biological sciences. The high separation efficiencies, short analysis times and great sensitivities (high analyte flux) associated with capillary zone electrophoresis (CZE) have also generated a significant amount of interest in this technique by an increasingly large number of analytical chemists, including mass spectrometrists. Since separation is effected quite differently in CZE than in HPLC, a brief description of the technique will be given below.

6.1 GENERAL PRINCIPLES

CZE separations are based on the differential migration of charged analytes that occurs in an electric field (1-6). This phenomenon has been widely recognized and utilized for a number of years by biologists and biochemists for the separation of biopolymers. Jorgenson and Lukacs (3) investigated the use of open capillary tubes to study zone-broadening in electrophoresis and found that capillary tubes offered a practical approach to rapid electrophoretic separations.

A schematic of a basic CZE system is shown in Figure 6.1. A fused silica capillary column with an inner diameter of 10-75 μm is placed so that one end is in a buffer solution at ground potential while the other end is placed in a buffer solution that is held at a high voltage potential. A typical operating voltage is 30 kV, but voltages of 10 to 50 kV are commonly used. The high-voltage power supply is connected to the CZE column through a high-voltage relay system. This relay (Model H-25, Kilovac Corp.) is used not only for high-voltage switching of the CZE system, but is also an integral part of the safety interlock system. For operator safety, the switching voltage of the relay is connected through a microswitch attached to the plexiglass box enclosing the high-voltage line, buffer solution, and CZE column (Figure 6.2). A second

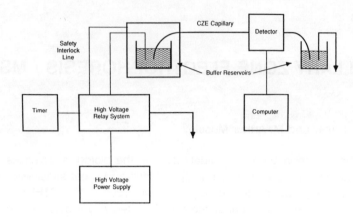

Figure 6.1. Schematic of a basic CZE system. (Reprinted with permission from reference 15.)

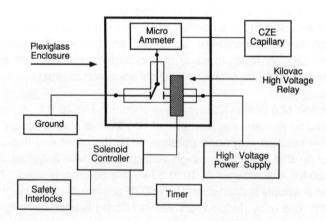

Figure 6.2. CZE high voltage control. (Reprinted with permission from reference 15.)

microswitch, attached to the plexiglass box, enables the high-voltage power supply. Thus, if the lid of the plexiglass box is opened, the power supply is disabled and the high-voltage relay disconnects the power supply from the CZE electrode, forcing the electrode to ground potential.

Migration of analytes between the two buffer reservoirs is effected by two mechanisms; electromigration and electro-osmosis. Electromigration occurs when a positively charged species migrates from the anode (high positive potential) to the cathode (low potential). The migration time is given by the equation: $t=L^2/mV$ where t is the migration time, L is the length of the capillary tube, m is the solute's electrophoretic velocity in a unit electric field and V is the applied voltage (6). L and V are the parameters that can readily be varied. The separation efficiency, N, is equal to $V/2D$ where D is the diffusion coefficient of the solute, assuming diffusion limited zone broadening. From this equation, one can see that a short tube with a high applied voltage will give the shortest migration times with the best separation. The limitation to this is that a significant amount of Joule heat will be generated which must be dissipated, otherwise zone broadening will occur due to the formation of a radial temperature profile within the CZE column. Either decreasing the column radius or increasing the column length will help in dissipating the Joule heat generated within the CZE capillary. Typical column lengths are in the range of 25 to 75 cm.

Electro-osmosis is the liquid flow occurring when an electrical potential is applied across a liquid-filled porous medium (7). The electro-osmotic flow acts to sweep all solutes through the capillary and does not promote separation. This can be envisaged as occurring by the movement of highly solvated cations toward the ground electrode, with this solvent flow moving other dissolved compounds and ions as well at rates of nanoliters per minute. Thus, even negatively charged analytes will be moved towards the ground electrode from the positively charged buffer solution, facilitating the analysis of both anions and cations in a single analysis.

Although CZE is not a chromatographic technique, it is sufficiently similar that many chromatographic concepts have been borrowed, such as migration time (retention time), resolution and theoretical plates. A major difference between chromatography and CZE is the difference in the flow profiles of the two techniques. Instead of the familiar parabolic laminar flow observed in chromatography (Figure 6.3(a)), a piston-like profile, with a constant velocity over most of the tube cross-section dropping to zero only near the walls, is observed with electro-osmosis (Figure 6.3(b)). A significant part of the superior separation efficiency of CZE over LC is due to the fact that the electro-osmotically driven, flat flow profile does not contribute to zone broadening whereas a pressure-driven parabolic flow profile is a substantial cause of this broadening.

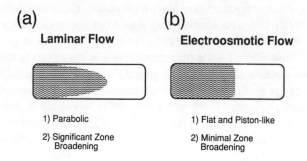

Figure 6.3. (a) Parabolic flow profile characteristic of chromatography. (b) Piston-like flow profile obtained from electro-osmotic flow.

Sample injection onto the CZE column can be effected by either of two means: electromigration sample introduction or hydrostatic sample introduction (8). In an electromigration sample introduction, the high voltage is turned off, the end of the capillary column is placed into a vial containing the sample (in a buffer solution identical to the CZE buffer or in water) and voltage is briefly applied across the CZE column. This causes the sample to be introduced onto the column via two distinctly different methods. First, the application of the voltage induces an electro-osmotic flow of sample solution into the column. Also, the application of the voltage causes electromigration of analyte ions. It should be noted that sample discrimination can occur between anions and cations due to the difference in their electromigration directions. After this sample loading step, the capillary is returned to the buffer reservoir and high voltage is again applied. Timing of the injection voltage is often controlled by a timer integrated into the high-voltage relay system (Figure 6.2). It should be noted that the safety interlocks and plexiglass housing are necessary to protect the operator from accidental contact with high voltage during this process. In hydrostatic sample introductions, the capillary end is placed into the sample vial and a pressure differential is applied to the column. After a calculated injection time, the pressure differential is removed and the capillary is placed into the buffer reservoir and voltage is applied. There is no discrimination between anions and cations when hydrostatic sample introductions are performed.

6.2 CAPILLARY ZONE ELECTROPHORESIS/MASS SPECTROMETRY

Interfacing CZE with mass spectrometric detection is, at present, based either on electrospray ionization (9-11), ion-spray (12, 13), or continuous-flow (CF) FAB (14-18). As the focus of this book is on CF-FAB, this chapter will be restricted to the interfacing of CZE with this technique. Within CZE/CF-FAB, two approaches have been used, the liquid junction interface (14, 17, 18) and the coaxial interface (15, 16).

6.2.1 Liquid Junction Interface

The liquid junction interface was first described by Minard (14) and has been utilized by van der Greef (18) and Caprioli (17) in various modifications. In the liquid junction interface (Figure 6.4), the CZE capillary terminates in a buffer

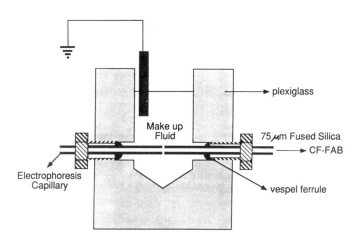

Figure 6.4. Liquid junction interface. (Reprinted with permission from reference 18.)

solution which contains the FAB matrix and is at ground potential. The end of the CF-FAB transfer line is placed in very close proximity to the end of the CZE capillary. A mixture of CZE effluent and the CF-FAB matrix solution are pulled into the CF-FAB transfer line (typically 50 μm (i.d.), 75 cm long) by the vacuum of the mass spectrometer. Since the vacuum-induced flow is approximately 5 μl/min and the CZE flow is of the order of 0.1 μl/min, efficient transfer of the CZE effluent into the transfer line occurs. Some zone broadening takes place using the liquid junction interface due to mixing effects. For example, Reinhoud *et al.* (18) observed a ten-fold loss of plate number and an accompanying three-fold loss in resolution in passing through the junction, compared with UV detection. A portion of this zone broadening can also be attributed to diffusion effects within the liquid film of FAB matrix on the probe tip which can increase the mean life-time of the analyte on the probe tip.

6.2.2 Coaxial Interface

The coaxial CZE/CF-FAB interface is an extension of the coaxial interface that was developed for interfacing 10 μm (i.d.) open tubular liquid chromatography and nanoscale packed capillary LC (50 μm (i.d.) packed columns) with CF-FAB (16, 19, 20, 21). This interface is shown schematically in Figure 6.5. The

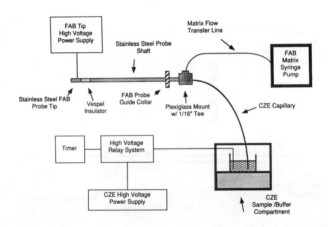

Figure 6.5. Schematic of the coaxial CZE/CF-FAB interface. (Reprinted with permission from reference 15.)

high-voltage circuitry and relay as well as the sample/buffer reservoir compartment are the same as with the off-line CZE apparatus. The "ground" end of the CZE capillary is located at the FAB probe tip (Figure 6.6(a)) with the FAB matrix flow being the "ground" reservoir. The sheath column should fit snugly into the hole in the probe tip to prevent backflow of the FAB matrix into the probe. Placement of a septum, teflon tape or Torr Seal in the probe tip also helps prevent backflow. The CZE capillary enters into the sheath column at the exterior end of the CF-FAB probe through a 1/16 inch Swagelock stainless steel tee (Figure 6.6(b)) with Vespel ferrules and terminates near the

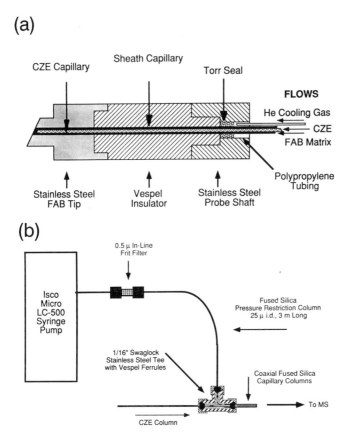

Figure 6.6. (a) Coaxial flow CZE-CF-FAB probe tip (reprinted with permission from reference 15.) (b) FAB matrix delivery system.

probe tip (about 1-2 mm within the sheath column). The CZE capillary is withdrawn slightly into the sheath column to reduce evaporation of the CZE buffer solution within the CZE capillary by the mass spectrometer's vacuum. Although the CZE capillary can be threaded through the sheath column from either direction, in practice, threading from the tee to the tip is preferred. We have noted that high-voltage arcing between the inside and outside of the CZE capillary (leading to formation of a hole) occurs more readily when threaded from tip to tee. The tee is mounted in a plexiglass handle to prevent electrical contact between the tee and the mass spectrometer source housing.

The ability to deliver the CZE effluent and the FAB matrix independently to the target is an important characteristic of the coaxial interface for several reasons. First, the volumetric flow rate of the glycerol solution that is required to maintain a stable ion beam is an order of magnitude or more greater than the total electro-osmotic flow rate of the buffer in the CZE system. Therefore, an independent means of delivering the glycerol matrix solution to the target is necessary. Second, the CZE column terminates at the target (FAB probe tip), which is held at +8 kV and constitutes the "ground" electrode of the CZE system. This facilitates active electrophoretic transport of the analytes directly to the FAB probe tip where ion desorption takes place. In addition, this design eliminates the need for a transfer line from the end of the CZE capillary to the FAB probe tip. Thus, it prevents zone broadening that would occur within such a transfer line, particularly if pressure-driven flow is used to transport the CZE separated analytes through the transfer line, and in the junction between the transfer line and the CZE column. The ability to independently deliver the two solutions to the target permits the independent optimization of the composition and control of the flow rates. The CZE buffer composition and pH can be optimized for the separation whereas the FAB matrix composition and pH can be optimized for analyte detection. This particular feature of the coaxial interface has proven to be particularly useful when ions are separated as negative ions but detected as positive ions (or vice versa).

Typical operating conditions for the coaxial CZE/CF-FAB interface are given in Table 6.1. As with liquid chromatography, all solutions should be scrupulously degassed and filtered, and the system should have no air

Table 6.1. Capillary Zone Electrophoresis Parameters

CZE Column = 13 μm i.d/150 μm o.d.
CZE Column Flow Rate = 20 - 50 nl/min
Sample Injection Volume = 0.25 - 5 nl
Buffer Composition = 0.005 M Ammonium Acetate adjusted to pH 8 or 8.5 with Ammonium Hydroxide
Effective Voltage Drop = 22 - 42 kV
Electro-osmotic Flow Rate = 30 nl/min
Sheath Column = 160 μm i.d./365 μm o.d.
FAB Matrix Composition = 25/75 Glycerol/Water with 0.5 mM Heptafluorobutyric Acid
FAB Matrix Flow Rate = 1 μl/min

pockets. A CZE column having an inner diameter of about 13 µm is used to prevent excessive vacuum-induced flow through the column. This is important in order to retain the high separation efficiency of CZE. Recall that electro-osmotic flow in CZE has a flat, piston-like flow profile across the capillary tube cross-section, dropping off only at the walls of the capillary (which does not significantly contribute to zone broadening). Pressure-driven flow, whether due to the use of high pressure on the inlet end of a column, as occurs in HPLC, or to low pressure on the outlet end of the column, due to the pressure differential between the mass spectrometer source and the CZE system, results in the formation of a parabolic flow profile across the capillary tube cross-section which in turn causes substantial zone broadening. Any pressure-driven flow, therefore, in the CZE column or in any of the transfer lines from the CZE column to the FAB probe tip will lead to zone broadening and, thus, a loss of CZE separation efficiency.

The FAB matrix solution used in our CZE interface contains heptafluorobutyric acid to aid in the detection of positive ions (matrix pH=3.5), or ammonium hydroxide for negative ions (matrix pH=9). These additives not only adjust the pH of the FAB matrix, but also ensure electrical contact between the end of the CZE column and the FAB probe tip by increasing the electrical conductivity of the FAB matrix solutions. The flow rate of the FAB matrix is controlled by a syringe pump (Figure 6.6(b)) and the volumetric flow rate is set to balance the rate of evaporation of the liquid from the probe tip. Improper flow rates are readily determined by an observed lack of ions derived from the matrix (too little flow) or by peak tailing (too much flow). The matrix solution transfer line from the pump to the tee incorporates a 0.5 µm in-line frit filter and a 3 m x 25 µm (i.d.) fused silica restriction column to provide sufficient back-pressure to the syringe pump for stable operation.

Using a 30 kV power supply, the effective voltage drop is 22 kV because the electrical "ground" is actually the +8 kV of the mass spectrometer source. This was a limitation of the CZE power supply initially used in our experiments. Recently, we have used a 60 kV power supply at potential of up to 50 kV, giving CZE separation potentials of up to 42 kV. The present limitation of 50 kV is due to the voltage limitation of the high-voltage relay.

Non-volatile buffers, such as potassium phosphate, create problems when used in conjunction with the coaxial interface. Sensitivity is reduced since the molecular ion species becomes divided between various cationized species such as $(M+H)^+$, $(M+K)^+$, $(M-H+2K)^+$, etc. (16). The buffer concentration used with the CZE/MS system is significantly lower than that usually employed in CZE. Buffer concentrations of 0.1 M to 0.01 M are generally used in CZE for two reasons. First, a high buffer concentration insures that the pH of the sample solution does not effect the buffer solution. Second, and most importantly, use of high buffer concentrations gives high conductivities and, therefore, high separation efficiencies. If the concentration of the buffer is not at least one hundred-fold higher than the analyte concentration, significant zone broadening will occur (3). Unfortunately, the use of buffer concentrations greater than 0.005 M of moderately volatile buffers, such as ammonium acetate, may lead to the formation of a "donut"-shaped spot on the FAB probe target tip. In this process, either radiation damage to the buffer forms a non-wettable surface surrounded by a ring of matrix on the probe tip, or evaporation of the aqueous matrix solution leads to formation of a "crust" which is surrounded by a ring of matrix. The net result is a dry surface exposed to the FAB beam with greatly reduced ion formation. This condition is readily detected because ion intensities fall suddenly and considerably.

The sample injection techniques which can be used for the coaxial interface are the same as for the other CZE/MS interfaces, i.e., hydrostatic (pressure-induced) and electromigration, but with some modifications. The hydrostatic injection is typically induced by the vacuum of the MS source, with no external pressure being applied to the introduction end of the CZE column. During the hydrostatic injections, care should be taken to insure that the FAB tip voltage is turned off or that the sample vial is not grounded. Otherwise, the positive voltage at the probe tip (for positive ion FAB) will induce an electro-osmotic flow from the FAB tip to the grounded capillary end, as well as a discrimination between anions and cations due to electromigration effects. An additional problem that has been noticed with electromigration sample introduction is an undesirable side effect of the rapid voltage changes due to high-voltage switching of the relay used in the CZE circuitry, presumably due to the generation of a radiofrequency electromagnetic pulse. If the mass spectrometer is scanning under computer control, the data system may lock-up, necessitating rebooting the computer system. Other computers in the vicinity of a CZE set-up may also be affected when the high-voltage switching occurs. These problems do not happen if the voltage of the CZE power supply is <10 kV during switching.

The coaxial interface preserves, to a large extent, the resolution inherent in CZE. The only contribution to zone broadening is by diffusion into the matrix at the probe tip, a source of peak broadening also noted with LC/CF-FAB. An example of the number of theoretical plates that can be generated using the coaxial CZE/CF-FAB technique is the electrophoretic separation of 250 fmol of Ala-Leu and 210 fmol of Asp-Leu in which 600,000 theoretical plates was achieved (16). Another example is shown in Figure 6.7 where two chemotactic peptides, Met-Leu-Phe and N-acetyl-Met-Leu-Phe, are readily separated by CZE/MS under negative ion conditions. This shows that minor changes in molecular structure readily affect the electrophoretic mobility and, therefore, lead to adequate separation analytes. Good sensitivity is again demonstrated with 46 fmol of analyte being readily observed in the negative ion mode.

Capillary Zone Electrophoresis/MS

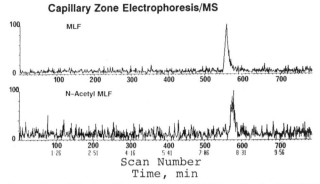

Figure 6.7. Separation of Met-Leu-Phe and N-acetyl-Met-Leu-Phe by CZE/CF-FAB MS.

6.2.3 Combination Liquid Junction/Coaxial Interface

Caprioli and coworkers (17) have developed a CZE/CF-FAB interface incorporating features of both the liquid junction and the coaxial approaches. Shown schematically in Figure 6.8, the liquid junction is contained in a 0.5 mm

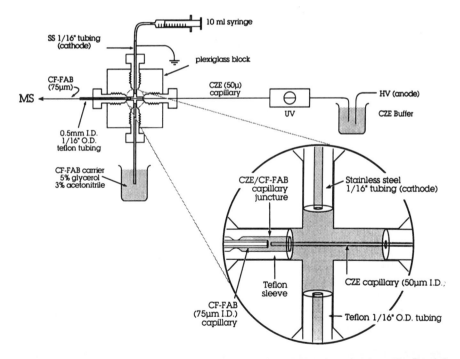

Figure 6.8. Schematic of combination liquid junction/coaxial interface developed by Caprioli and coworkers. (Reprinted with permission from reference 17.)

(i.d.) PTFE tube and the CZE capillary may be placed at any distance from, as well as inside, the CF-FAB capillary. In effect, this interface is an adjustable coaxial design, allowing the analyst to adjust parameters for a particular instrument arrangement or analytical requirement.

Although a number of operating parameters affect separation efficiencies, relatively high salt concentrations in the CZE buffer system are generally recommended for best performance. Unfortunately, FAB ionization does not perform well under such conditions and the analysis is characterized by poor sensitivity and the appearance of sodium adduct ions in the spectra. This is shown in Figure 6.9 for a peptide of $(M+H)^+$ m/z 674 in the CZE/CF-FAB

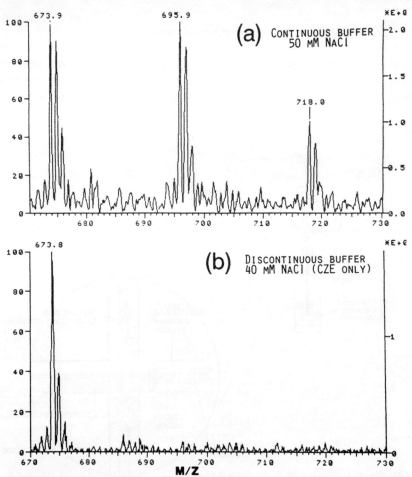

Figure 6.9. Mass spectra of the molecular ion region of a peptide in a tryptic digest analyzed by CZE/CF-FAB using continuous and non-continuous buffer systems. Details are given in the text. (Reprinted with permission from reference 17.)

analysis of the tryptic digest of β-lactoglobulin (17). Figure 6.9(a) shows the spectrum taken from the analysis using a "continuous" buffer system, i.e., both the CZE and CF-FAB (liquid junction) solutions contained 50 mM NaCl. The spectrum in Figure 6.9(b) was obtained using a "discontinuous" electrophoresis buffer system, i.e., the CZE buffer contained 40 mM NaCl, but the liquid junction reservoir contained 3 mM ammonium acetate and not alkali salts at set-up. It is seen that the discontinuous buffer gives the best CF-FAB performance in producing only $(M+H)^+$ and thereby gives optimal sensitivity since the total ion intensity resides in a single molecular species.

The application of CZE/CF-FAB MS for the analysis of tryptic digests of proteins is, at the moment, of prime interest. An example of this is shown in Figure 6.10 for UV (a) and MS (b) detectors used simultaneously for the

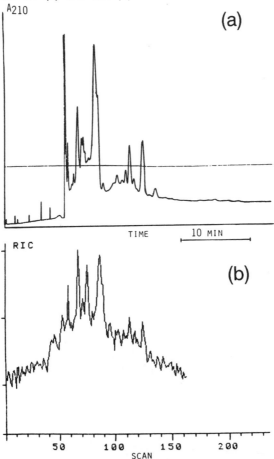

Figure 6.10. Analysis of the tryptic digest of approximately 40 pmol of recombinant human growth hormone using (a) UV and (b) MS detection in tandem. Details are given in the text. (Reprinted with permission from reference 17.)

analysis of the tryptic digest of recombinant human growth hormone (17). The total ion chromatogram represents the sum of the entire range scanned by the mass spectrometer (m/z 400-2000) and therefore in itself is relatively insensitive. However, specific ion chromatograms, shown in Figure 6.11 for four peptides, illustrates both the sensitivity and specificity of the CZE/MS combination.

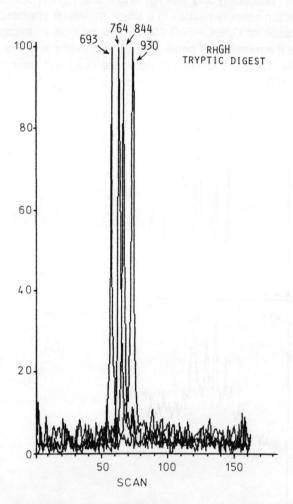

Figure 6.11. Selected ion chromatograms of four tryptic peptides in the CZE/CF-FAB analysis of human growth hormone shown in Figure 6.9. The $(M+H)^+$ value measured for each compound is labeled at the top of the peak.

High salt concentrations in the CZE buffer solution can be used with this liquid junction interface while still employing low salt concentrations in the matrix solution. In effect, the higher flow rate of the matrix solution sufficiently dilutes the salt concentration from the CZE capillary so that the cationized molecule ion species due to salt are of negligible relative abundance.

6.2.4 Advantages and Limitations

The major advantage of CZE/CF-FAB is the ability to detect the $(M+H)^+$ molecular species of a wide variety of polar and charged molecules at high sensitivities. This is particularly useful for the analysis of complex mixtures.

A disadvantage of the use of FAB is the relatively high chemical background recorded, especially at low masses ($<m/z$ 400). Because there have been no direct comparisons of different ionization techniques for the same compound using identical instruments and operating parameters, it is not possible at this point to determine whether or not the chemical background is a significant problem relative to other techniques. For example, we have analyzed 350 fmol of 2,4-D with negative ion CZE/CF-FAB and observed a signal-to-noise ratio of approximately 2.5:1 for $(M+H)^+$ with a scan range of 70 daltons. This compares to a report showing a signal-to-noise ratio of 20:1 for 1.2 pmol of this compound using CZE/electrospray MS with single-ion monitoring (13). The sensitivities are roughly comparable, with increased signal-to-noise observed with electrospray over CF-FAB in these studies attributed to the increased amount of analyte injected and the use of single-ion monitoring instead of scanning techniques for data collection.

Another disadvantage is that one works with high-voltage equipment and must take care to insure that the proper safety interlocks are in place. Finally, as in all capillary liquid techniques, proper plumbing connections, careful attention to eliminating dead volumes, and the use of filtered and degassed solutions are essential to efficient operating conditions.

6.3 CONCLUSION

Two different systems have been used for interfacing CZE with continuous-flow FAB - the liquid junction and the coaxial column configuration. These interfaces provide simple and efficient methods of using CF-FAB/mass spectrometry as a detector for capillary zone electrophoresis. This combination yields a very powerful analytical technique which is especially well suited for the analysis of ionic, thermally labile compounds and, therefore, is especially useful in the analysis of biomolecules.

REFERENCES

1. F.E.P. Mikkers, F.M. Everaerts, and T.P.E.M. Verheggen, *J. Chromatogr. 169*, 11 (1979).
2. J.W. Jorgenson, and K.D. Lukacs, *Clin. Chem. 27*, 1551 (1981).
3. J.W. Jorgenson, and K.D. Lukacs, *Anal. Chem.* 53, 1298 (1981).
4. J.W. Jorgenson, and K.D. Lukacs, *J. High Res. Chromatogr. Chromatogr. Commun. 4*, 230, (1981).
5. J.W. Jorgenson, and K.D. Lukacs, *Anal. Chem. 53*, 1298 (1981).
6. J.W. Jorgenson, and K.D. Lukacs, *Science 222*, 266 (1983).
7. C.L. Rice, and R. Whitehead, *J. Phys. Chem. 69*, 4017 (1965).
8. D.J. Rose,Jr., and J.W. Jorgenson, *Anal. Chem. 60*, 1840 (1988).
9. J.A. Olivares, N.T. Nguyen, C.R. Yonker, and R.D. Smith, *Anal. Chem. 59,* 1230 (1987).
10. R.D. Smith, C.J. Baringa, and H.R. Udseth, *Anal. Chem. 60*, 1948 (1988).
11. R.D. Smith, J.A. Olivares, N.T. Nguyen, and H.R. Udseth, *Anal. Chem. 60*, 436, (1988).
12. E.D. Lee, W. Mueck, J.D. Henion, and T.R. Covey, *J.Chromatogr. 458*, 313 (1988).
13. E.D. Lee, W. Mueck, J.D. Henion, and T.R. Covey, *Biomed. Environ. Mass Spectrom. 18*, 253 (1989).
14. R.D. Minard, D. Chin-Fatt, P. Curry, and A.G. Ewing, *Proceedings of the 36th Annual Conference on Mass Spectrometry and Allied Topics*, San Francisco, CA, 5-10 June 1988, 950.
15. M.A. Moseley, L.J. Deterding, K.B. Tomer, and J.W. Jorgenson, *Rapid Commun. Mass Spectrom. 3*, 87 (1989).
16. M.A. Moseley, L.J. Deterding, K.B. Tomer, and J.W. Jorgenson, *J. Chromatogr. 480*, 197 (1989).
17. R.M. Caprioli, W.T. Moore, M. Martin, B.B. DaGue, K. Wilson, and S. Moring, *J. Chromatogr. 480*, 247 (1989).
18. N.J. Reinhoud, W.M.A. Niessen, U.R. Tjaden, L.G. Gramberg, E.R. Verheij, and J.v.d. Greef, *Rapid Commun. Mass Spectrom. 3*, 348 (1989).
19. J.S.M. deWit, L.J. Deterding, M.A. Moseley, K.B. Tomer, and J.W. Jorgenson, *Rapid Commun. Mass Spectrom. 2*, 100 (1988).
20. M.A. Moseley, L. Deterding, K. Tomer, R.T. Kennedy, N.L. Bragg, and J.W. Jorgenson, *Anal. Chem. 61*, 1577 (1989).
21. L.J. Deterding, M.A. Moseley, K.B. Tomer, and J.W. Jorgenson, *Anal. Chem. 61*, 2504 (1989).

Chapter 7

ANALYSIS OF LOW-POLARITY SUBSTANCES

David L. Smith

Fast atom bombardment mass spectrometry has been used to analyze substances with a wide range of polarities. Although procedures for assessing polar substances such as peptides and nucleotides are highly developed, methods for analyzing less polar substances have received little attention. This is due in part to the relatively poor performance of FAB for analyzing these substances. FAB mass spectra of nonpolar compounds are often characterized by weak molecular ion signals and a high level of background due to the matrix. In addition, electron and chemical ionization have been relatively successful for many low-polarity substances, thus reducing the need for FAB analysis.

Several laboratories have demonstrated the improved performance of CF-FAB MS for different types of compounds (1-3). Advantages of this sample introduction method include reduced background and uniform sensitivity for mixtures of substances with a range of polarities. This chapter will deal with the use of CF-FAB MS for the analysis of substances with low to intermediate polarity. Examples are presented for the analysis of two members of a class of low-polarity natural products, acetogenins (4,5), and metabolites of benzo[a]pyrene (6,7). Several examples will be given for the analysis of completely non-aqueous samples. In addition, the design of CF-FAB probes for use on Kratos MS 25 and MS 50 mass spectrometers will be discussed.

7.1 PROBE DESIGN

The purpose of the CF-FAB probe is to provide a flow of carrier liquid to the FAB ion source where sample ions are desorbed as the liquid flows across the probe tip. Controlled evaporation of the liquid flowing across the probe tip, which is essential for stable operation, is achieved by heating the tip so that the rate of evaporation just balances the rate of liquid delivery. Maintaining a stable liquid film is difficult because water, as well as most other frequently used carrier liquids, is in a state of boiling in the high-vacuum chamber. As a result, water tends to form bubbles which disrupt the flow. Good stability is achieved when the carrier liquid flows as a thin film across the probe tip. This is a most fortunate consequence since many of the advantages of CF-FAB MS may be traced to ion desorption from a thin film of liquid.

General features of the CF-FAB probes used in our laboratory are illustrated in Figure 7.1. The shaft is attached to a stainless steel probe end via a Vespel insulator. The probe end is threaded for a standard 10-32 stainless steel screw, which serves as a removable probe tip. The center of the tip is drilled to accept an insert which consists of a length of standard 1/16 inch (o.d.) stainless steel tubing. This design avoids problems associated with drilling long, small-diameter holes in stainless steel, and provides a convenient means of adjusting the rate of heat transfer to the liquid inside the fused silica capillary. Evacuation holes, located in the Vespel insulator and the probe end piece, are used to remove air during insertion, and vapors from the carrier liquid which may be located on the exterior of the fused silica capillary during operation. A septum is included as a safety device to prevent possible migration of the liquid along the exterior of the fused capillary. This back-flow of carrier liquid can lead to instability of the liquid sample surface through vaporization and can also generate a path of low resistance between the high voltage in the ion source and ground. This aspect is discussed in more detail in Chapter 1.

Features considered important to the design of a CF-FAB probe include ease of assembly, temperature control of the tip, and shape and orientation of the tip surface from which the ions are sampled. For example, gravitational force plays a role in directing, and perhaps maintaining, the flow of the carrier fluid. Although both the MS 25 and MS 50 have vertical geometries, the angle between the ion beam as it is accelerated in the ion source and gravitational force (F_G) are different. In the MS 50, this angle is 135°, and in the MS 25, it is 0°. For instruments with horizontal geometry, this angle is 90°. The design of the CF-FAB probe tips used for the two instruments is shown in Figure 7.1.

The rate of solvent evaporation is critically dependent on the temperature of the probe tip, which is heated by physical contact with a temperature-controlled source heating block. For a flow of 5 μl/min of water, the heat flow from the source heating block to the tip must be approximately 12 J/min if the temperature of the tip is to remain unchanged. Only 6 J/min is required for a similar flow of methanol. Although the optimum temperature of the tip is usually determined empirically, poor contact between the source heating block and the end of the probe can lead to unstable operation because of poor temperature control. It is not practical to determine the temperature of the surface of the tip from which vaporization occurs. However, we estimate, from the geometry of the probe and the thermal conductivity of

Analysis of Low-Polarity Substances

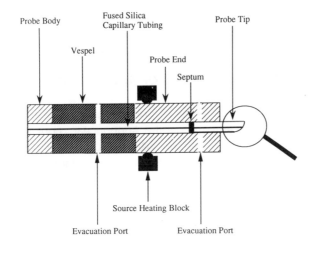

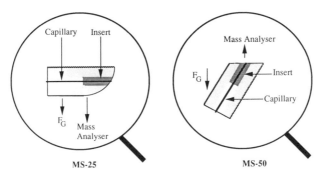

Figure 7.1. CF-FAB probe design used on the Kratos MS-25 and MS-50 mass spectrometers. The magnified portion of the figure shows the probe tips, with the Earth's gravitational force F_G, indicated for each tip when located in the ion source.

stainless steel, that the temperature drop from the block to the probe tip is less than 5 °C.

Heat transfer from the tip through the fused silica capillary to the solvent is also an important parameter. If this process is highly efficient, the solvent can be heated above its boiling temperature and bubbling inside the capillary may occur. If heat transfer through the capillary is poor, ice may form at the tip of the capillary from the water in the carrier solution. Heat transfer to the capillary depends principally on the wall thickness of the capillary, the contact between the fused silica capillary and the tip, and the temperature of the tip. For the probe design illustrated in Figure 7.1, heat transfer to the solvent can also be varied by using stainless steel inserts of different lengths and having holes of varying inner diameters.

7.2 ANALYSIS OF ACETOGENINS

Acetogenins, which are a series of polyketide-derived fatty acid derivatives found in the bark of certain trees, the Annonaceae, contain tetrahydrofuran rings and a methylated γ-lactone with hydroxyl, acetoxyl, and/or ketoxyl groups along the hydrocarbon chain (4,5). Various members of the acetogenins have potent biological activities, such as cytotoxic, antitumor, antimalarial, antimicrobial, immunosuppressant, and pesticidal activity. Structures of two acetogenins, bullatalicin and bullatacin, are given in Figure 7.2. Because the acetogenins often differ by only the placement of a hydroxyl

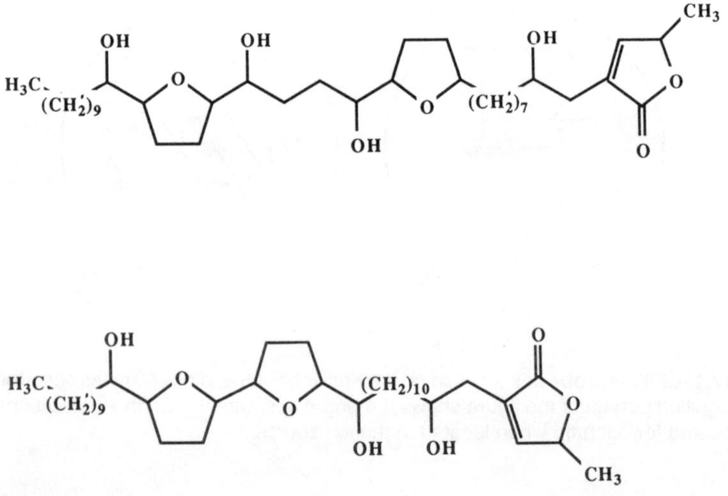

Figure 7.2. Structures of representative acetogenins: (a) bullatalicin and (b) bullatacin.

group along the hydrocarbon chain, it is important to have highly selective analytical methods for distinguishing among the different acetogenins. In addition, it is important to have sensitive and highly specific analytical methods for quantifying acetogenins in crude extracts.

Several features of acetogenins combine to confound their structure elucidation. Being insoluble in water, the acetogenins are generally characterized as nonpolar. However, the electron ionization mass spectra of acetogenins usually have few peaks from which their structures can be reliably predicted. As a result, structure elucidation strategies have relied heavily on analyses of acetyl and trimethylsilyl derivatives (8,9). Conventional FAB mass spectra of acetogenins have a high level of background and no structural information. The isolated material often contains more than one component because the members of the acetogenin family have similar physical and chemical properties. HPLC has been used sparingly for isolating acetogenins because their absorbance above 215 nm is low, making their detection difficult. Because of these problems, the structures of some acetogenins have been reported incorrectly (4).

Although EI mass spectra of the trimethylsilyl derivatives of acetogenins are rich in structural information, CF-FAB MS is of interest because it can be used to analyze acetogenins directly. This feature is important because of the possibility that structural changes may occur during derivatization. In addition, CF-FAB MS offers the capability of using on-line HPLC/MS to survey complex extracts for specific, biologically active acetogenins. CF-FAB may also be combined with MS/MS to give important structural information.

FAB mass spectrometry has been most successful for analyses of highly polar substances because of their propensity to form ions in solution, or to become charged as they are sputtered at the liquid/vacuum interface (10,11). Success in the positive ion mode is related to the fact that polar analytes are stronger bases (i.e., they have greater proton affinities) than the matrix. Complementary logic has been used to develop strategies for analysis by negative ion FAB MS. Nonpolar substances are difficult to analyze by FABMS because they are neutral, and because they have low proton affinities.

Since nonpolar substances are concentrated in the surface layers of polar matrices at the liquid/vacuum interface, attempts to optimize CF-FAB MS for analysis of acetogenins have focused primarily on finding conditions that produce charged species. A preliminary investigation of the effect of adding acid (formic or trifluoroacetic acid) or cation-forming species (sodium iodide) showed that these techniques fail to enhance the intensity of molecular ions. However, excellent spectra were obtained when the carrier liquid (methanol, chloroform, or acetonitrile) contained a low concentration (1%) of thioglycerol, or dithiothreitol and dithioerythritol (DTT/DTE) in addition to 3% glycerol. Satisfactory spectra could not be obtained when m-nitrobenzylalcohol (NBA) was mixed with the carrier liquid. However, good spectra were achieved when any of these matrices (DTT/DTE, thioglycerol, or NBA) were mixed with the sample and injected into the carrier fluid (methanol) which contained glycerol. The CF-FAB mass spectrum for direct injection of bullatacin (Figure 7.3) was taken using this procedure. This spectrum has an intense peak at m/z 623, which corresponds to the molecular ion $(M+H)^+$. Peaks at m/z 551, 569, 587, and 605 are due to multiple losses of water from the molecular ion. The background level is also much lower than that in normal FAB. This example

CF-FAB Mass Spectrometry

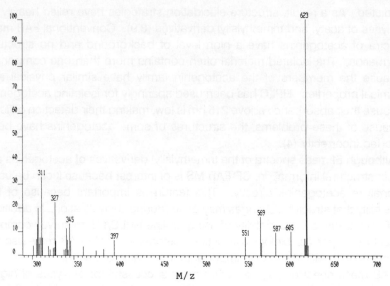

Figure 7.3. CF-FAB mass spectrum of bullatacin. Five hundred nanograms were dissolved in a 50:50 mixture of methanol and NBA and injected into a carrier liquid composed of acetonitrile, methanol and glycerol (50:45:5). The flow rate was 6 μl/min. Scans taken on either side of the peak were used to reduce the background in this spectrum.

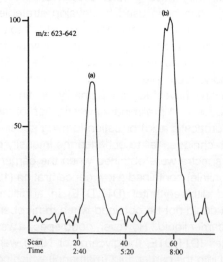

Figure 7.4. Plot of the sum of ion signals from m/z 623 to m/z 642 obtained for the reversed phase HPLC separation of bullatacin (a) and bullatalicin (b). The mobile phase consisted of methanol/water/glycerol/thioglycerol (87:7:2:4). A gradient HPLC system was used to generate a flow of 30 μl/min. Separation was achieved isocratically with an Adsorbosphere C18 HS 5 μm, 1 mm x 25 cm column. The effluent from the column was split 4:1 to give a flow into the ion source of 6 μl/min.

illustrates the utility of CF-FAB in studies to find the optimum matrix for a sample by allowing co-injection of matrix compounds with the sample.

We have also used CF-FAB MS to monitor on-line the HPLC reversed-phase separation of acetogenins. A plot of the intensity in the molecular ion region versus time (Figure 7.4) shows that acetogenins can be separated by reversed-phase HPLC and detected by CF-FAB MS. The mass spectrum of bullatalicin (Figure 7.5) has a prominent molecular ion (M+H)$^+$ at m/z 639.

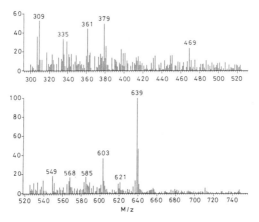

Figure 7.5. CF-FAB mass spectrum of bullatalicin taken as it elutes from the reversed-phase column. See Figure 7.4 for experimental details.

Other intense peaks can be attributed to fragment ions diagnostic of the structure of bullatalicin. These results demonstrate that direct coupling of HPLC and CF-FAB MS can be useful for rapid screening of plant extracts for specific acetogenins. The next steps in this investigation will focus on optimizing conditions for screening such extracts and the use of MS/MS analysis of the molecular ions to obtain additional structural information.

7.3 ANALYSIS OF METABOLITES OF BENZO[A]PYRENE

The carcinogen benzo[a]pyrene is metabolically converted to carcinogenic derivatives prior to binding to macromolecules and inducing cancer (6,7). The study of the multiple pathways of metabolism of benzo[a]pyrene is thus essential to understanding the mechanism of carcinogenesis. The metabolites of benzo[a]pyrene include phenols, diols, tetrols, and water-soluble glucuronic acid, sulfate and glutathione conjugates. Identification of the conjugates by electron impact or chemical ionization mass spectrometry has been difficult because these metabolites are involatile and thermally unstable (12). Although

FAB mass spectrometry has proven useful for analyzing a wide variety of involatile or thermally labile substances, detection limits are poor because of the characteristically high level of background signals. For this reason, we have investigated the potential of CF-FAB MS as a sensitive and highly selective means of identifying metabolites of benzo[a]pyrene.

Glucuronic acid and sulfate conjugates of benzo[a]pyrene give excellent negative ion FAB mass spectra because they form anions easily. Initial attempts to operate the Kratos MS-50 mass spectrometer at 8 kV failed because of excessive arcing between the metal rods used to fix the beam-centering plates and the grounded source slit assembly. Removal of a set of 1 mm quartz spacers on the slit side of these plates significantly reduced the frequency of arcing. This modification does not change the spacing of the plates and hence does not affect the focusing properties of the acceleration lens assembly.

The negative ion standard FAB mass spectrum of 2.5 µg of glucuronic acid conjugate (Figure 7.6(a)) has a peak corresponding to the molecular (M-H)⁻ ion

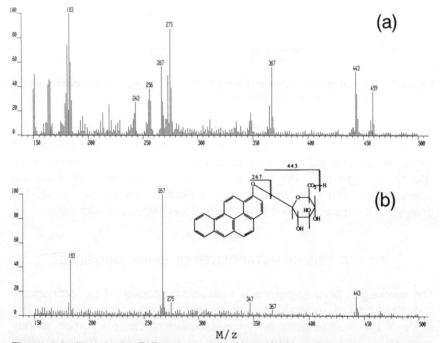

Figure 7.6. Negative ion FAB mass spectra of benzo[a]pyreneβ-d-glucopyranosiduronic acid obtained using (a) standard FAB (2.5 µg) and (b) CF-FAB MS (0.1 µg). The CF-FAB mass spectrum was taken with a flow of 3 µl/min of methanol/water/glycerol (10:10:1) using a syringe pump.

at m/z 443, with intensity comparable to the intensities of the glycerol background peaks (m/z 275, 367, and 459). Fragment peaks diagnostic for the structure of the metabolite cannot be readily distinguished from the background. In contrast, results obtained with CF-FAB for 25 times less material (Figure 7.6(b)) gave an intense molecular ion species as well as a number of peaks that can be correlated to the structure of the compound. These results demonstrate that CF-FAB provides a substantial improvement in signal-to-noise ratio and detection limit over standard FAB for analysis of the benzo[a]pyrene glucuronide conjugate. Similar results have also been obtained for the sulfate conjugate (B[a]p-SO_4).

Although the background level is reduced in CF-FAB, initial attempts at achieving a detection limit in the low nanogram range were not successful. However, it was observed that an instrument mass resolution of 3,000 was sufficient to resolve the B[a]p-SO_4 ions from the glycerol background. Although the source pressure associated with CF-FAB MS is high, the peaks were only slightly broadened at the base. The separation of B[a]p-SO_4 ions from background ions in CF-FAB and standard FAB is illustrated in Figure 7.7. These peaks were generated using repetitive narrow scans over a mass window centered at m/z 347, the $(M+H)^+$ ion of B[a]p-SO_4. Results from two scans are included to illustrate their reproducibility. The mass peak for the molecular ion of B[a]p-SO_4 is indicated in the figure by shading. Although the background signal from glycerol is relatively small for 15 ng of B[a]p-SO_4, it becomes increasingly important for smaller quantities. The results of these studies indicate that the limit of detection of B[a]p-SO_4 by high-resolution CF-FAB MS is approximately 0.5 ng. For standard FAB MS, the background peak at m/z 347 is about four times greater than the B[a]p-SO_4 peak, whereas for CF-FAB, the B[a]p-SO_4 peak is about seven times greater than the background peak. These results demonstrate the significant improvement in signal-to-background when CF-FAB is used. Given the differences in the peak heights and recorder sensitivities for the two modes of operation, the CF-FAB signal is about 20 times greater than for standard FAB at the 15 ng level for this compound.

With the addition of a reversed-phase HPLC column, CF-FAB MS can be used to analyze mixtures of metabolites of benzo[a]pyrene, as illustrated in Figure 7.8. A mixture containing 100 ng of B[a]p-SO_4 and 400 ng of B[a]p-glucuronide was analyzed by selected ion monitoring of a fragment ion, m/z 267, which is common to both metabolites. A C-18 (5 μm) 1 mm x 25 cm column was used with gradient elution from 75% (methanol/water/glycerol) to 100% in 25 minutes at a flow rate of 30 μl/min. The column effluent was split 5:1, allowing a flow rate of 5 μl/min to pass into the mass spectrometer.

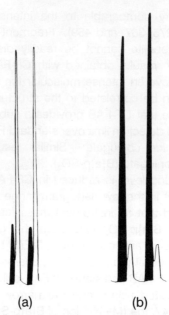

Figure 7.7. Repetitive scans at a resolution of 3,000 over a narrow mass window centered at m/z 347. Fifteen nanograms of benzo[a]pyrene-sulfate were used for both the standard FAB (a) and the CF-FAB (b) measurements. The benzo[a]pyrene-sulfate molecular ion is indicated by shading. The signal intensity is 1 volt full scale in panel (a), and 5 volts full scale in (b). For CF-FAB, the solvent (water/methanol/glycerol, 10:10:1) flow rate was 3 µl/min.

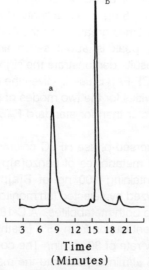

Figure 7.8. Single ion recording (m/z 267) of on-line LC/CF-FAB MS analysis of a mixture of benzo[a]pyrene-3-sulfate (peak a) and benzo[a]pyrene-β-d-pyranosideglucuronic acid (peak b).

7.4 OTHER NON-AQUEOUS APPLICATIONS

Barber *et al.* (13) reported the analysis of several water-insoluble compounds including fats, lipids and glycerides with CF-FAB MS by using organic solvents alone, or a mixture of these with water, as the carrier solvent. Samples were introduced by flow-injection and were dissolved in a suitable organic solvent. Compounds such as tripalmitin and 1,2- and 1,3- distearin did not show an $(M+H)^+$, but produced intense ions for the loss of a fatty acid side chain and the elimination of water. Intense fragment ions were also produced by further side chain losses. The carrier solvent for these samples was acetonitrile containing 5% thioglycerol. These authors also reported the spectrum of cholesterol obtained with CF-FAB using a 1:1 water/acetonitrile mixture containing 5% glycerol. The sample of cholesterol was dissolved in chloroform. The mass spectrum shows an intense $(M-H)^+$ molecular ion species at *m/z* 385, and a major fragment at *m/z* 369, as shown in Figure 7.9.

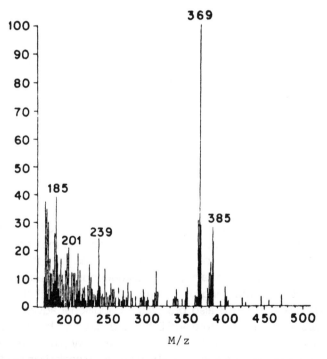

Figure 7.9. The CF-FAB mass spectrum of cholesterol dissolved in chloroform. (Reprinted with permission from reference 13.)

Studies have also been reported by Barber and coworkers (14) for the analysis of air- and moisture-sensitive manganese/phosphine compounds. These are of interest because of their ability to bind gas molecules such as oxygen in a manner that resembles that of heme-containing globins. Previous attempts using standard FAB for the analysis of analogous phosphine oxide compounds were unsuccessful. For the CF-FAB analysis of tertiary butyl phosphines, dry toluene was used as the carrier solvent. The mass spectrum shows intense ions for the M^+ species, with little or no oxidation evident. In contrast, standard FAB analysis, even with very limited exposure to air through use of a fume hood and syringe sampling from sealed systems, showed a predominant oxide molecular ion species.

Organic solvents in high concentrations are used with LC/CF-FAB in the analyses of mixtures of compounds by reversed phase chromatographic techniques. Solvent systems containing 50-70% or more of acetonitrile, methanol, or other solvent are used isocratically or in gradient elution programs. A number of these applications have been described in Chapter 5, LC/MS, which have used acetonitrile as the eluting solvent, a common system for the elution of peptides from C-8 and C-18 columns. Boulenguer *et al.* (15) have used a solvent consisting of methanol:thioglycerol:water (90:10:5, by volume) for the elution of permethylated oligosaccharides from a C-18 packed capillary column. The gradient consisted of 80% to 100% of this solvent at a flow rate of 5 μl/min. Similarly, Barefoot *et al.* (16) have used a solvent consisting of methanol:water:glycerol (80:10:10, by volume) for the separation and analysis of the plant herbicides sulfometuron methyl and chlorinuron ethyl and their metabolites in plants.

7.5 CONCLUSION

The analysis of nonpolar compounds by CF-FAB using non-aqueous solvent systems as the carrier solution provides an easy and effective means of obtaining mass spectra. The composition of solvent and the concentration of compound remain constant during the analysis and sample redistribution effects are minimized. For certain types of compounds such as lipids, fats and oils, solvent systems employing chloroform, toluene, and methanol have been successfully used. CF-FAB also provides an additional advantage in some special cases where air, moisture and light are destructive to a sample. Since the technique can be used with dry solvents, with oxygen nearly completely purged from the system, samples may be taken from sealed vials with colored glass or opaque syringes and directly flow-injected into the mass spectrometer with no exposure to the environment.

ACKNOWLEDGMENTS

The author thanks Geraldine Thevenon and Yohannes Teffera for their help in obtaining data presented here and preparation of the manuscript, and J. McLaughlin and W. Baird, who provided reference materials for the acetogenins and benzo[a]pyrene conjugates, respectively. The author also thanks the National Institutes of Health for financial support through grant GM RO1 40384.

REFERENCES

1. R.M. Caprioli, and T. Fan, *Biochem. Biophys. Res. 141*, 1058-1065, (1986a).
2. R.M. Caprioli, and T. Fan, *Anal. Chem. 58*, 2951-2954, (1986b).
3. J.A. Page, and M.T. Beer, *J. Chromatog. 474*, 51-58, (1989).
4. J.K. Rupprecht, Y-h. Hui, and J. McLaughlin, *J. Nat. Prod. 53*, (1990).
5. T.T. Dabrah, and A.T. Sneden, *J. Nat. Prod. 47*, 652-657, (1984).
6. R.G. Harvey, *Acc. Chem. Res. 14*, 218-226, (1981).
7. A. Dipple, *Chemical Carcinogens*, edited by C.E. Searle, American Chemical Society Monograph, 173, 245-314, 1976.
8. A. Alkofahi, J.K. Rupprecht, D.L. Smith, C-j. Chang, and J. McLaughlin, *Experientia 44*, 83-85, 1988.
9. S.D. Jolad, J.J. Hoffmann, K.H. Schram, and J.R. Cole, *J. Org. Chem. 47*, 3151-3153 (1982).
10. S.J. Pachuta, and R.G. Cooks, *Chem. Rev. 87*, 647-669 (1987).
11. J. Sunner, A. Morales, and P. Kebarle, *Anal. Chem. 59*, 1378-1383 (1987).
12. R.H. Bieri, and J. Greaves, *Biomed. Environ. Mass Spectr. 14*, 555-561 (1987).
13. M. Barber, L.W. Tetler, D. Bell, A.E. Ashcroft, R.S. Brown, and C. Moore, *Org. Mass Spectrometry 22*, 647 (1987).
14. M. Barber, D.J. Bell, M. Morris, L.W. Tetler, M. Woods, G.A. Gott, P.P. MacRory, and C.A. McAuliff, *Proceedings of the 36th ASMS Conference on Mass Spectrometry and Allied Topics*, San Francisco, CA, 5-10 June 1988, 545.
15. P. Boulenguer, Y. Leroy, J.M. Alonso, J. Montreuil, and B. Fournet, *Anal. Biochem. 168*, 164 (1988).
16. A.C. Barefoot, R.W. Riser, and S.A. Cousins, *J. Chromatogr. 474*, 39 (1989).

Chapter 8

OTHER APPLICATIONS

8.1 DEVELOPMENT OF CF-FAB MASS SPECTROMETRY FOR THE BIOPHARMACEUTICAL INDUSTRY

James H. Bourell, William J. Henzel, John Frenz, William S. Hancock and John T. Stults

The biopharmaceutical industry, in its efforts to discover and produce protein pharmaceuticals, requires an extensive amount of protein characterization throughout a product's development. Often, in the discovery phase, only small amounts of native (human) material are available for analysis. Thus, sensitive protein characterization techniques are required. Stringent regulatory guidelines for pharmaceutical production require that formulated proteins be fully characterized for product integrity. Here, the ability to characterize small amounts of possible process-related modifications within product formulations is important. HPLC and, more recently, mass spectrometry have become essential to the accomplishment of these characterizations. The linking of HPLC to continuous-flow fast atom bombardment mass spectrometry (CF-FAB MS) has been developed to augment these analytical efforts by increasing sensitivity and decreasing analysis time.

We have made considerable progress in developing a sensitive and highly reproducible HPLC/CF-FAB MS system. It employs a packed capillary HPLC column operated at a flow rate of 3 to 5 μl/min, which allows the entire effluent to pass into the mass spectrometer through a fritted CF-FAB probe interface. Thus, the amount of sample required for analysis is reduced. The detection limit is at the subpicomole level. The chromatographic resolution is comparable to that obtained on 4.6 mm diameter columns. Resolution is not seriously diminished upon transfer to the mass spectrometer. These characteristics permit HPLC/CF-FAB MS to be used to identify peptides in HPLC maps of protein digests.

As an additional approach to identifying trace amounts of modified proteins in product formulations, we have coupled high-performance displacement chromatography (HPDC) (1) with CF-FAB MS. Under displacement conditions, samples 50 to 100 times the loading capacity found in conventional chromatography are injected on the column. Trace components within the sample elute as highly concentrated peaks between displacement bands of major components. Using a packed capillary column allows HPDC to be performed at flow rates amenable to the entire sample being introduced to the MS, thereby maximizing effective trace level sensitivity.

In this section we wish to discuss the key aspects of our work. More detailed experimental procedures and their results have been published recently (2,3).

8.1.1 Packed Capillary HPLC of Peptides

HPLC with packed capillary columns, while requiring more careful assembly of components, provide excellent separations with lower solvent consumption. The narrow column diameter results in peak elution volumes of 2 to 4 μl. Thus, the concentration of each peak is higher than that obtained with full-sized or microbore columns. Optimal flow rates are between 3 and 5 μl/min. Similar resolution is achieved to that obtained with conventional columns at much higher flow rates. The low flow rates exactly match the optimal flow rate for the CF-FAB MS interface. The columns we use are sold commercially (LC Packings). They consist of a 320 μm by 15 cm fused silica capillary packed with 3 μm particle size, C-18 reverse-phase material. The performance of these columns has been excellent. With proper care, over 50 injections on the same column is possible.

The transfer volume, after the column, should be kept as low as possible to minimize peak broadening in the UV detector and the mass spectrometer. For this reason, only capillary tubing (50 μm diameter) with zero dead-volume connections is used after the column. Approximately 150 cm of tubing, about 3 μl total volume, is required to connect the column to the mass spectrometer and to provide in-line UV detection.

The separation of peptides by reverse-phase HPLC is an adsorption/desorption process. Peptides are adsorbed and remain bound to the column until a critical percentage of organic modifier desorbs and elutes the peptide (4). Thus, injection volumes and dead volumes found before the column are not as critical as they are for HPLC separations based on partitioning. Figure 8.1 shows that resolution and sensitivity are maintained even when a large volume (5 μl) injection loop is used. We have found that a 5 μl injection loop offers a reasonably small volume while still allowing enough volume to accommodate dilute injections. However, a relatively large volume between the mixing tee and the front of the column leads to a long time delay for gradient conditions to arrive at the column. Low flow rates

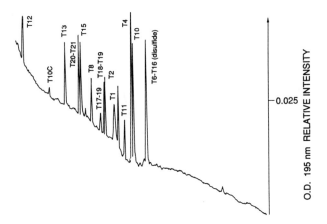

Figure 8.1. UV absorbance trace for a 5 picomole (in 5 μl) injection of human growth hormone (met-hGH) tryptic peptides run on the system described in Figure 8.2. HPLC conditions: water/glycerol/TFA [99/1/0.1 (solvent A)], acetonitrile/glycerol/TFA [99/1/0.1 (solvent B)], with a gradient program of 2% to 60% B in 65 minutes after an initial hold time 10 minutes, and 195 nm UV detection. (Reprinted with permission from reference 2.)

further lengthen this delay.

We have found UV detection with the capillary flow cell to be invaluable for the development of the HPLC/CF-FAB MS system. It is used to make sure HPLC separations are taking place properly before the components enter the mass spectrometer. Also, valuable MS instrument time is saved by using the UV detector for HPLC method development. The capillary flow cell consists of a sapphire bead focusing lens mounted in the light path of a holder for the cleared section of capillary tubing. The path length of the cell, determined by the diameter of capillary tubing, is very small, but the expected loss in sensitivity is more than compensated for by the high concentration of solutes eluting from the packed capillary column. Because of the high solute concentration and the low solvent absorption, due to the short light path, we are able to detect at 195 nm wavelength. This gives approximately five times more sensitivity than operating at 214 nm with the same cell. Detection limits for peptides of about ten residues in length is near the picomole level.

8.1.2 CF-FAB Mass Spectrometry

A schematic diagram of the HPLC/CF-FAB MS probe is shown in Figure 8.2. In the design of the Frit-FAB interface (JEOL, LTD) the capillary tubing passes through an insulated probe support and makes contact with a thin microporus frit (5). The frit is supported by a metal cover that screws into the probe and

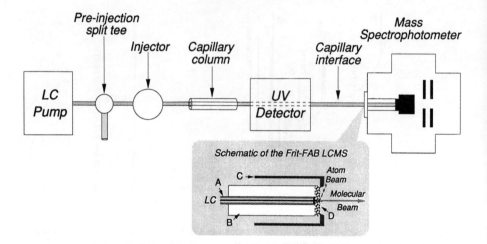

Figure 8.2. Schematic diagram of the Frit-FAB HPLC MS system. The system consists of a ABI microgradient pump, ABI model 783 UV detector equipped with the CZE on-column cell, and a 0.32 x 150 mm C18 column from LC Packings. A preinjection split (approx. 20 to 1) is used to allow optimum gradient mixing conditions at 100 μl/min. Inset: Detail of the Frit-FAB target: (A) fused silica capillary, (B) support guide, (C) cap, and (D) stainless steel frit. (Reprinted with permission from reference 2.)

has been drilled out in such a way as to provide differential pumping towards the outer rim of the frit and away from the source. Analytes emerge from the tubing with the solvents and matrix and pass through the frit to the target surface which is exposed to the ionization beam. The matrix, with dissolved analytes, continues to flow towards the outer rim of the frit disc. In this way the target area clears rapidly for the next eluted analyte. The UV and extracted ion profiles for the same peptide in a typical HPLC/CF-FAB MS experiment show the extracted ion peaks to be broadened by about 40% in comparison to the UV peak (see Figure 8.3), which is a predicted result of the capillary interface volume. We observe almost no tailing due to the frit not being cleared effectively. The Frit-FAB probe, like other types of CF probes, requires an elevated temperature (45 to 50 °C) to stabilize the ion current.

The total ion current (TIC) trace for the tryptic map of recombinant human growth hormone is shown in Figure 8.3. We can identify almost all of the expected peptides at the 25 pmol level. Figure 8.4 shows the reconstructed ion current profile (a) and the mass spectrum (b) for the amino terminal tryptic fragment from a 0.5 pmol injection of the hGH digest. Even at this level peptides smaller than 1500 u gave interpretable mass spectra and all, with the exception of the disulfide linked T6-T16 peptide ($[M+H]^+ = m/z$ 3763), gave significant peaks in their reconstructed ion current profile.

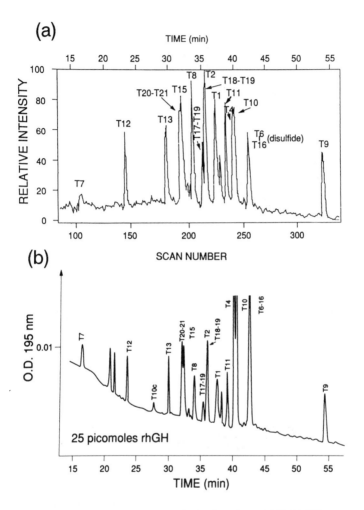

Figure 8.3. The tryptic maps from a 25 pmol injection of met-hGH: (a) TIC trace, and (b) 195 nm UV trace. HPLC conditions are the same as in Figure 8.1. (Reprinted with permission from reference 2.)

Enzymatic digest maps of recombinant proteins are not usually sample limited and almost always are developed on 4.6 mm diameter columns with high flow rates. The HPLC/CF-FAB MS analysis for these proteins can be done by splitting the flow after the column (5,6).

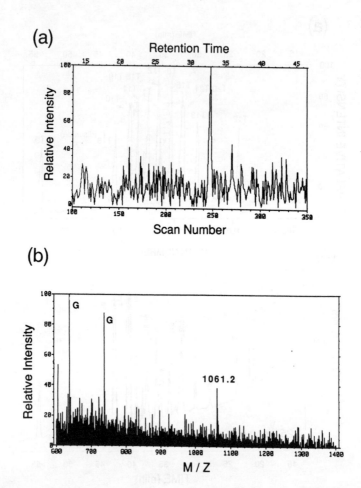

Figure 8.4. (a) The reconstructed ion current profile for *m/z* 1061-1062 (tryptic peptide T1) from a 500 fmol injection of trypsin digested met-hGH. (b) The mass spectrum from scan number 247. The peaks labelled "G" are protonated glycerol oligomers. (Reprinted with permission from reference 2.)

8.1.3 Displacement Chromatography/CF-FAB MS

Our latest efforts to characterize trace impurities found in enzymatic digests have led us to investigate linking HPDC with CF-FAB MS. A comparison of the UV trace and the TIC trace for an HPDC separation of trypsin digested hGH is shown in Figure 8.5. Ten nanomoles of digested material were loaded

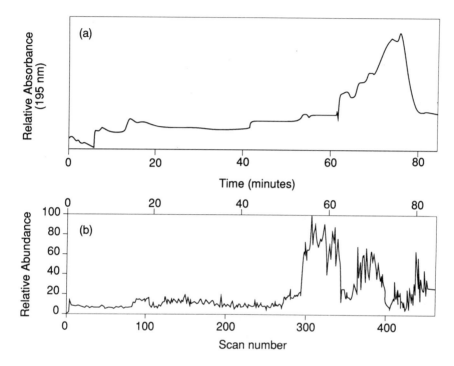

Figure 8.5. The profiles obtained during the displacement chromatography of 10 nmol of digested met-hGH monitored by (a) absorbance of light at 195 nm and (b) the total ion current in the mass spectrometer. Spectra were acquired in 12-second scans. (Reprinted with permission from reference 3.)

on the column. Peptides elute from the column in zones dependent upon the size of the column, the amounts of peptide and displacer eluting off the column, and many other factors (1). As shown in Figure 8.6, the mass spectrometer offers a definitive way to identify these zones. Minor peptide components become sandwiched in highly concentrated low-volume peaks between the major peptides that show true displacement behavior. This is shown in Figure 8.7 where the scans between two major peptide zones, T10c1 (m/z 537) and T12 (m/z 773), contain peptides with m/z values of 604, 620, 659, 755, and 805. These trace fragments are thought to be derived from nonspecific cleavages of hGH but further sequence analysis is required to confirm their structures.

The high concentrations required for HPDC present problems of frit clogging and shortened column life. These problems may be solved by changes in matrix concentration and solvent composition.

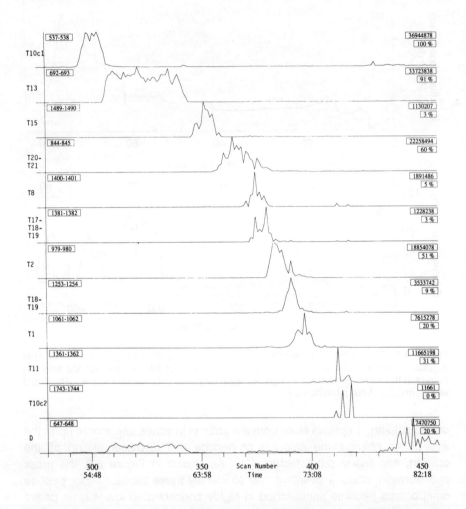

Figure 8.6. Reconstructed ion current chromatograms for the displacement chromatogram shown in Figure 8.5. The ions shown correspond to the predominant peptides produced from the tryptic digest of met-hGH (see reference 2 for sequence information and exact masses of the common tryptic fragments). The boxes in the left corner of each trace give the nominal mass range counted. The boxes in the right corner correspond to total area counts and their relative area percentage. (Reprinted with permission from reference 3.)

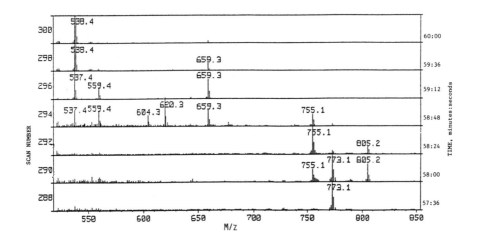

Figure 8.7. The mass spectra, for scans 288 through 301, taken from data acquired during the displacement run shown in Figure 8.5. Each spectrum represents the sum of two scans (i.e., scan 288 is actually 288+289). (Reprinted with permission from reference 3.)

8.1.4 Conclusion

We have found HPLC/CF-FAB MS well suited for direct analysis of proteins available in sparingly small amounts. This approach to mass spectral analysis, using packed capillary columns, parallels our efforts to improve sensitivity in other areas of protein characterization such as amino acid analysis and protein sequencing. The Frit-FAB probe gives results comparable to those obtained with other types of CF probes (7,8). Moreover, linking HPDC to mass spectrometry, we have been able to demonstrate trace component enhancement and have been able to better define displacement behavior for individual peptides. Further research is required to establish the practicality of this method.

REFERENCES

1. J. Frenz, and Cs. Horvath, in *HPLC-Advances and Perspectives Vol. 5*, edited by Cs. Horvath, Academic Press, New York, p. 212 (1989).
2. W.J. Henzel, J.H. Bourell, J.T. Stults, *Anal. Biochem.,187*, 228 (190).
3. J. Frenz, J.H. Bourell, and W.S. Hancock, *J. Chromatogr.* In press.
4. W.S. Hancock, and J.T. Sparow, In *HPLC-Advances and Perspectives Vol. 3*, edited by Cs. Horvath, Academic Press, New York, p. 49 (1983).
5. D. Ishii, T. Takeuchi, *Trends in Anal. Chem.*, 8, 25 (1989).
6. R.M. Caprioli, B.B. DaGue, and K. Wilson, *J. Chromatogr. Sci.* 26, 640 (1988).
7. S.J. Gaskell, and R. Orkiszewski, Chapter 2, this volume.
8. R.M. Caprioli, B. DaGue, T. Fan, and W.T. Moore, *Biochem. Biophys. Res. Commun.* 146, 291 (1987).

8.2 BIOMEDICAL APPLICATIONS OF GRADIENT CAPILLARY LC/MS USING CF-FAB

Bradley L. Ackermann, John E. Coutant, and Teng-Man Chen

Over the past few years, LC/MS has become a very valuable technique to the biomedical laboratory. The power of LC/MS for such applications is that, as opposed to other methods, it enables direct examination of components in complex biological matrices. While several options for LC/MS currently exist, because of its sensitivity and scope of application, continuous-flow FAB (CF-FAB) has generated considerable interest (1-4). One constraint of the method, however, is the limitation on flow rate (e.g., <10 µl/min). This means that LC methods which employ greater flow rates, such as conventional bore (4.6 mm i.d.) and microbore (1 mm i.d.), must undergo an effluent split prior to the mass spectrometer and therefore limits the sensitivity of these methods. Since biomedical applications frequently involve sample limited situations, packed capillary LC provides a viable alternative as it is compatible with low microliter flow rates. Consequently, the entire sample injected on-column reaches the mass spectrometer.

Recently, we designed and implemented a system for capillary CF-FAB in our laboratory (5). A schematic overview of the system appears in Figure 8.8 and incorporates the following features: gradient elution, post-column matrix addition, and variable wavelength on-line UV detection. These features were combined with capillary LC to provide a LC/MS interface which is both efficient and practical. Gradient elution, for example, not only optimizes the separation

Other Applications

of complex mixtures, but also permits the use of large injection volumes (0.5 - 1.0 μl) with minimal band broadening. Further, because the system forms gradients at conventional flow rates (1.0 - 1.5 ml/min), without the FAB matrix in the mobile phase, it is capable of reproducing conditions developed using conventional bore HPLC. Post-column matrix addition offers several other advantages in terms of flexibility and avoids any adverse effects on the chromatography caused by having glycerol in the mobile phase. Moreover, the use of a coaxial flow interface has proven to be an effective means for post-column matrix addition as it does not yield any significant contribution to band broadening. Finally, on-line UV detection was incorporated into the system to permit the absorption profile to be observed during the analysis and because it does not involve a split of the effluent to the mass spectrometer.

To date, the described LC/MS interface has been used for several applications of biomedical interest. The data presented here represent two analyses where sensitivity was an important consideration and are examples where the biomolecules of interest were refractory to other methods of ionization.

8.2.1 Experimental

Gradients were formed by a conventional HPLC system at a flow rate of 1.5 μl/min. The mobile phase was then split before the injector using a Valco low-dead-volume tee to give a flow rate of about 2 μl/min. A Rheodyne model 7413 injection valve with 0.5 μl loop was used for sample introduction. The split ratio was determined dynamically by a conventional bore HPLC column placed in parallel with the capillary column as shown in Figure 8.8. The capillary column was slurry packed in-house with 5 μm Spherisorb-ODS using 320 μm (i.d.) fused silica (Polymicro Technologies). The end of the capillary column was interfaced to a fused silica transfer line, 50 μm (i.d.), 500 μm (o.d.), using a low-dead-volume tee. The two columns were held in proximity (1 mm) by a fused silica sheath 530 μm (i.d.). The FAB matrix solution glycerol:CH_3CN:H_2O, 1:2:1) was introduced into the tee by a syringe pump (Waters) at a flow rate of 1 μl/min and allowed to flow to coaxially around the end of the packed column. On-line UV detection was accomplished by removing a 1 cm segment of the polyimide coating from the transfer line and positioning this portion in the optical path of a Kratos Spectraflow 757 UV detector. The interface to the mass spectrometer (VG ZAB2-SE) was accomplished by threading the transfer line through the flow probe provided by the vendor. A flow rate of 3 μl/min was measured at the probe tip. An ion source temperature of about 40 °C was used for all analyses.

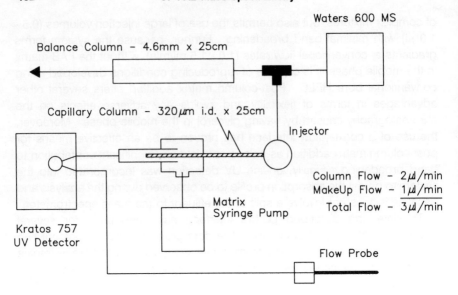

Figure 8.8. Diagram of the capillary LC system used for CF-FAB. (Reprinted with permission from reference 5.)

8.2.2 Analysis of Glycopeptide Antibiotic Mixtures

Teicoplanin is a glycopeptide antibiotic produced by *Actinoplanes teichomyceticus* and occurs as a complex mixture containing six major components. Each compound has a generalized structure consisting of a peptide backbone internalized as part of a multi-ring system to which two to three sugar residues are attached. Teicoplanin components are differentiated by unique fatty acid side chains incorporated into an N-acyl-D-glucosamine residue. Separation of teicoplanin mixtures may be obtained using reversed-phase HPLC where it has been shown that retention is governed by the fatty acid moieties (6).

Teicoplanin provides an excellent test case for any LC/MS system because, in addition to the chromatographic challenge, the molecules are difficult to ionize by other MS methods. Figure 8.9 displays the UV and total ion profiles for the analysis of 1 nmol teicoplanin by capillary LC/CF-FAB. A 50-minute linear gradient from 15-50% acetonitrile in water containing 0.1% TFA was used for the analysis. The data indicate good separation of the major components and that the chromatographic integrity was not compromised by the LC/MS interface. A mass spectrum obtained for the major component (A2-2) is shown in Figure 8.10. In addition to the $(M+H)^+$ ion at m/z 1880, other diagnostic fragment ions were observed at this concentration including a peak at m/z 1564 corresponding to the loss of the N-acyl-D-glucosamine residue.

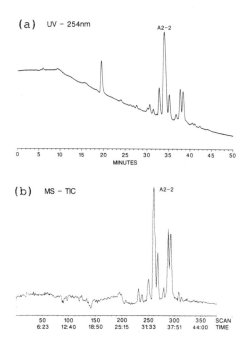

Figure 8.9. UV and total ion chromatograms for the gradient separation of 1 nmol of teicoplanin. (Reprinted with permission from reference 5.)

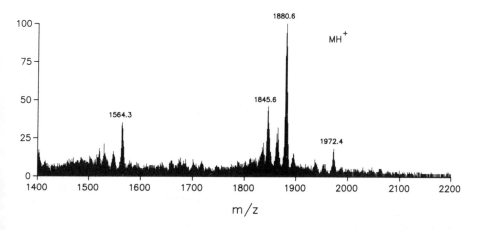

Figure 8.10. Mass spectrum of major teicoplanin component A2-2. (Reprinted with permission from reference 5.)

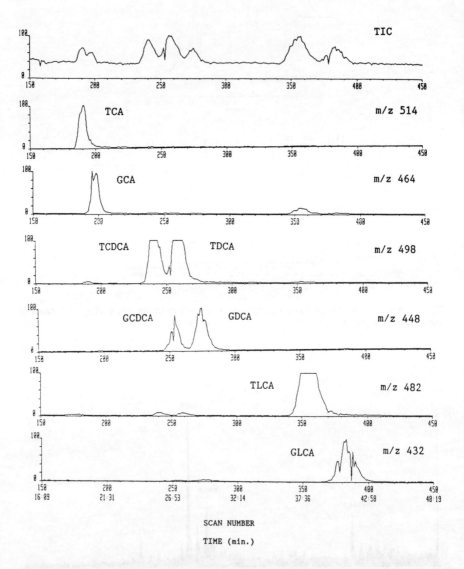

Figure 8.11. Total ion and individual mass chromatograms for eight endogenous bile acid standards (50 ng ea.) analyzed in the negative ion mode by capillary LC/CF-FAB. The observed order of elution was: taurocholic acid (TCA), glycocholic acid (GCA), taurochenodeoxycholic acid (TCDCA), glycochenodeoxycholic acid (GCDCA), taurodeoxycholic acid (TDCA), glycodeoxycholic acid (GDCA), taurolithocholic acid (TLCA), and glycolithocholic acid (GLCA).

8.2.3 Profiling of Bile Conjugates in Biological Fluids

Bile conjugates are important molecules having a wide range of biological function in both metabolic and hormonal systems. These molecules are also diagnostic, since changes in their relative profiles serve to indicate various disease states. Unfortunately, the lack of a chromaphore makes their detection difficult using HPLC (7). In addition, because these compounds are difficult to ionize by conventional methods, GC/MS is not a method of choice.

Initial studies performed in the negative ion mode using capillary CF-FAB have shown encouraging results for the separation and detection of these compounds. Figure 8.11 shows a series of mass chromatograms for the (M-H)⁻ ions of eight endogenous bile salt standards (50 ng each). These data were acquired under a gradient starting at (25:75) methanol/0.4 M ammonium acetate (pH 5.7) and ramped linearly over 3 minutes to a final composition of (75:25) methanol/0.4 M ammonium acetate (pH 5.7). Using these conditions, separation was achieved for all eight conjugates. Figure 8.12 shows the region

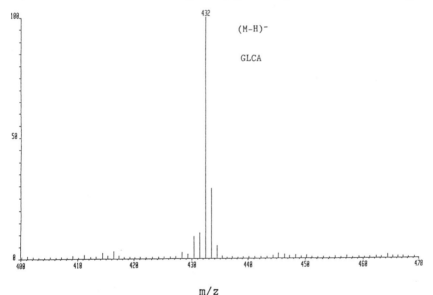

Figure 8.12. Region surrounding the (M-H)⁻ ion in the mass spectrum of glycolithocholic acid (MW 433).

surrounding the molecular ion (M-H)⁻ for the least sensitive component, glycolithocholic acid (GLCA). The signal-to-noise estimated from these data (>100:1) indicates that subnanogram quantities should be readily detected for bile conjugates in the full scanning mode. Work aimed at using capillary CF-FAB to profile bile conjugates in urine, serum, and bile is ongoing.

REFERENCES

1. R.M. Caprioli, T. Fan, and J.S. Cottrell, *Anal. Chem. 58,* 2249 (1986).
2. Y. Ito, T. Toyohide, D. Ishii, and M. Goto, *J. Chromatogr. 346,* 161 (1985).
3. R.M. Caprioli, B. DaGue, T. Fan, and W.T. Moore, *Biochem. Biophys. Res. Comm. 146,* 291 (1987).
4. D.E. Games, S. Pleasance, E.D. Ramsey, and M.A. McDowall, *Biomed. Environ. Mass Spectrom. 15,* 179 (1988).
5. J.E. Coutant, T-M Chen, and B.L. Ackermann, *J. of Chromatogr. 529,* 265 (1990).
6. A. Borghi, C. Coronelli, L. Faniuolo, G. Allievi, R. Pallanza, and G.G. Gallo, *J. Antibiot. 37,* 615 (1984).
7. J. Street and K.D.R. Setchell, *Biomed. Chromatogr. 2,* 229 (1988).

8.3 MICROBIAL BIOTRANSFORMATION AND MICROCOLUMN LC/CF-FAB MS FOR SULFONYLUREA HERBICIDE METABOLITE IDENTIFICATION

R. W. Reiser and B. Stieglitz

8.3.1 Summary

Microbial biotransformation of the sulfonylurea herbicides chlorsulfuron and chlorimuron ethyl were obtained with the soil bacterium *Streptomyces griseolus*. The major metabolites in crude broth extracts were rapidly identified using LC/CF-FAB MS. These compounds are extremely thermally labile, and cannot be analyzed by GC/MS and often do not give molecular ions using LC/MS with other ionization techniques. The FAB mass spectra gave the protonated molecular ion as base peak, and structurally useful fragment ions. Microbial metabolites are often the same as plant, animal, and soil metabolites, and their rapid generation and identification facilitate later environmental studies of these systems.

8.3.2 Introduction

Microbial biotransformation of pesticides provides a rapid method for generating metabolites, and LC/MS with FAB ionization provides a rapid technique for their identification. It has been found that *Streptomyces griseolus* efficiently metabolizes many sulfonylurea herbicides (1). Microbially derived metabolites can be isolated and purified for use as standards for comparison with metabolites obtained in soil, plants and animals. The biotransformation broth extracts are much cleaner than soil, plant or animal extracts, and can be analyzed directly without further clean-up.

In previous work, the use of packed fused silica capillary LC columns of 0.25 mm (i.d.) allowed us to obtain improved performance in reversed-phase LC/MS analyses with a moving belt interface (2-4). We attained high-quality EI and CI mass spectra on 0.1 μg samples injected into the LC column. However, FAB mass spectra are often needed to obtain unequivocal molecular weights of polar and/or thermally labile metabolites. LC/MS using FAB ionization with a moving belt interface has been reported in the literature (5, 6), but sensitivity is poor, possibly due to the short residence time of the sample in the ion source. With the continuous-flow (CF) FAB interface (7), the sample remains on the FAB target until it is ionized or vaporized, giving excellent sensitivity. The flow rates used with 0.25 mm (i.d.) packed capillary (1-2 μl/min) are ideally suited to direct interfacing with CF-FAB, i.e., no splitter or make-up flow are required.

Sulfonylurea (SU) herbicides and their metabolites are extremely thermally labile, and cannot be analyzed as the intact molecule by GC/MS due to their thermal degradation in the injection port or column. We have been unable to derivatize SU herbicides for GC/MS due to their decomposition during the derivatization process or thermal instability of the derivatives. Most SU herbicides and metabolites do not show significant molecular ions in their EI, CI or thermospray mass spectra. LC/MS with FAB ionization is an ideal technique for analyses of these labile compounds (4, 8). This section reports the identification of microbial metabolites of chlorsulfuron, the active ingredient in Glean® wheat herbicide, and chlorimuron ethyl, the active ingredient in Classic® soybean herbicide.

8.3.3 Experimental

The experimental protocol is shown in Figure 8.13. *Streptomyces griseolus* (ATCC 11796) was grown on sporulation medium for 24-48 hours to obtain a good biomass yield (growth phase). Sulfonylurea (SU) degrading enzymes were induced by adding 25-50 ppm of the SU under study and incubating for 4-24 hours (induction). The induced cells were incubated with 100-160 ppm SU in fresh sporulation medium for 4 or 24 hours, or in C-salts (chemically defined medium) for 4, 24 or 48 hours (biotransformation). The cells are removed by centrifugation and membrane filtration and the supernatant is

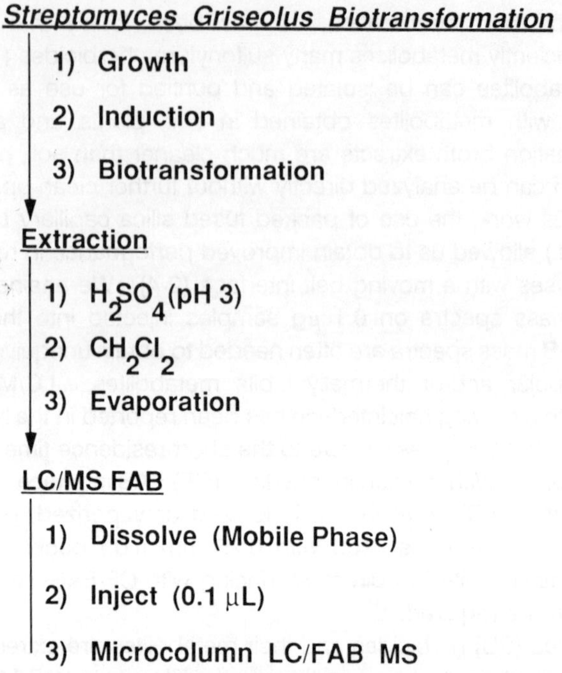

Streptomyces Griseolus Biotransformation

1) Growth
2) Induction
3) Biotransformation

Extraction

1) H_2SO_4 (pH 3)
2) CH_2Cl_2
3) Evaporation

LC/MS FAB

1) Dissolve (Mobile Phase)
2) Inject (0.1 μL)
3) Microcolumn LC/FAB MS

Figure 8.13. Experimental protocol outline.

acidified to pH 3 and extracted with two volumes of methylene chloride. The extract is evaporated to dryness and the residue is dissolved in 50/50 acetonitrile/water (pH 3, formic acid) at approximately 1 μg/μl, and 0.1 μl is injected for LC/CF-FAB MS.

The CF-FAB interface was designed to be interchangeable with our moving belt interface, as described previously (4, 8). A schematic diagram of the equipment is shown on Figure 8.14. The FAB target is heated by radiation from the ion source, and runs at 45-50 °C with the source at 200 °C. Heat is needed to prevent freezing of the LC solvent as it evaporates (7). FAB mass spectra obtained on SU herbicides and other thermally labile compounds using these conditions were identical to those achieved with the conventional FAB probe with the ion source at room temperature, demonstrating that the 200 °C source temperature does not induce thermal degradation. An Ion Tech FAB gun was used with xenon gas at 8 kV anode voltage.

The LC mobile phase was 50% acetonitrile/40% water (pH 3, formic acid)/10% glycerol. Addition of 10% glycerol to the mobile phase has caused no adverse affects to the LC pump or the chromatographic separations. The LC column was slurry packed with 3 μm Zorbax ODS, as previously described (2).

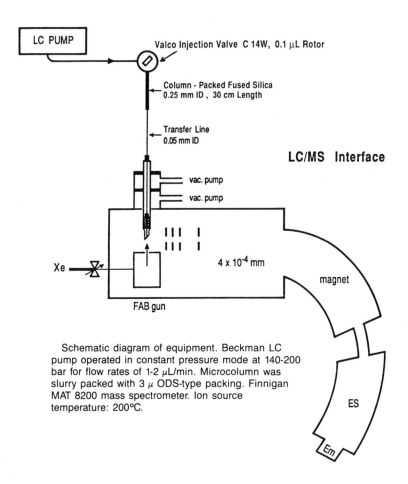

Figure 8.14. Schematic diagram of microcolumn LC/CF-FAB MS.

Figure 8.15 shows a schematic diagram of the threaded FAB target incorporating a Vespel/graphite ferrule to prevent backflow of liquid into the probe. The hole in the FAB target was drilled to give a snug fit with the fused silica capillary. The mass range scanned was m/z 45-650 at 2 s/decade scan rate. Glycerol matrix ions were readily eliminated from the spectra by subtracting the background from either side of the LC peak. The mass spectrometer resolution was 1,000. Mass calibration is conveniently checked using the glycerol cluster ions.

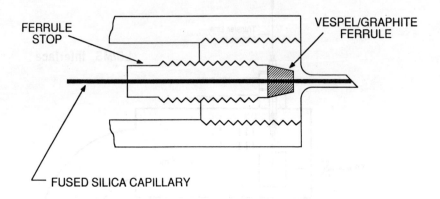

Figure 8.15. Schematic diagram of seal at FAB target.

8.3.4 Results and Discussion

The LC/UV chromatogram obtained off-line on the chlorsulfuron biotransformation broth extract is shown in Figure 8.16. Two major metabolites, both more polar than the parent compound, were produced. The CF-FAB mass spectra and structures of chlorsulfuron and its two microbial metabolites are shown in Figure 8.17. The glycerol matrix ions were subtracted out of all spectra. The protonated molecular ion is the base peak in the FAB spectra of all three components, and the isotopic clusters show all three are mono chloro compounds. The molecular weight of metabolite 1 is 14 mass units lower than the parent, and the mass of the protonated triazine amine fragment ion is also 14 below that obtained in the parent spectrum, indicating O-demethylation has occurred. The molecular weight of metabolite 2 is 16 mass units higher than the parent, and the triazine amine fragment is also 16 mass units higher than that obtained in the parent spectrum, indicating hydroxylation occurred on the triazine side of the molecule.

The LC separation obtained on the chlorimuron ethyl biotransformation broth extract is shown on Figure 8.18. The CF-FAB spectra (Figure 8.19) again show the protonated molecular ions as the base peak for all three

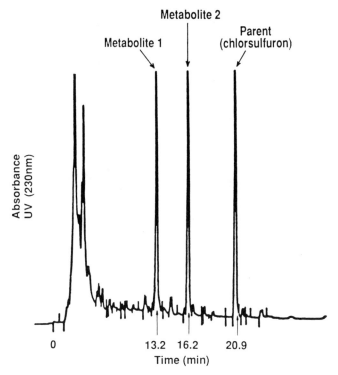

Figure 8.16. Liquid chromatogram of chlorsulfuron broth extract obtained using a Zorbax ODS column with a UV detector.

components, and the isotope pattern shows all three are monochloro compounds. The molecular weight of metabolite 1 is 14 mass units below that of the parent, and the mass of the protonated pyrimidine amine fragment ion is also 14 lower than that obtained in the parent spectrum, indicating metabolite 1 is the O-demethylated parent. Metabolite 2 has a molecular weight 28 mass units below that of the parent, but the mass of the pyrimidine amine fragment ion is the same as that of the parent. This shows the pyrimidine portion of the molecule is unchanged, and O-de-ethylation occurred on the benzene side.

In these studies, we were able to assign structures of the four metabolites from the FAB mass spectral data alone, since there was only one chemically logical structure that fit the data. In other cases, additional spectral data (NMR, IR, MS/MS) or synthesis are needed for complete structure determinations of metabolites.

CF-FAB Mass Spectrometry

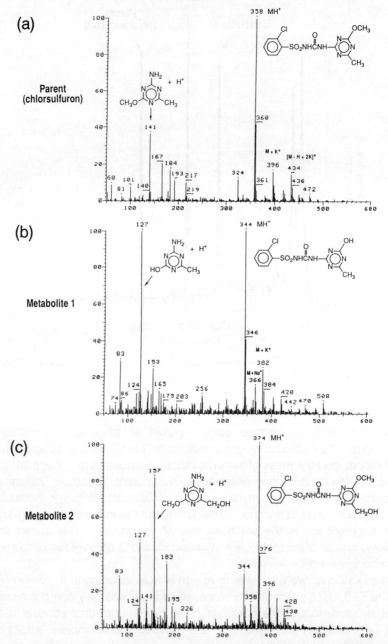

Figure 8.17. CF-FAB mass spectra of chlorsulfuron and microbial metabolites from an LC/MS analysis.

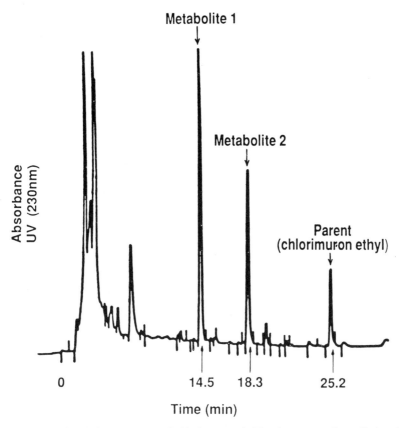

Figure 8.18. Liquid chromatogram of chlorimuron ethyl broth extract using a Zorbax ODS column and UV detector.

8.3.5 Conclusions

Coupling microbial bioconversion with LC/MS using CF-FAB has allowed us to rapidly generate and identify sulfonylurea herbicide metabolites. Broth extracts were analyzed directly, with no additional clean-up. FAB mass spectra have provided unequivocal molecular weights and structurally useful fragment ions of these extremely thermally labile compounds. This information allows preparation of metabolite standards, either synthetically or microbially, for use in metabolism and residue studies required by government agencies for registration of herbicides.

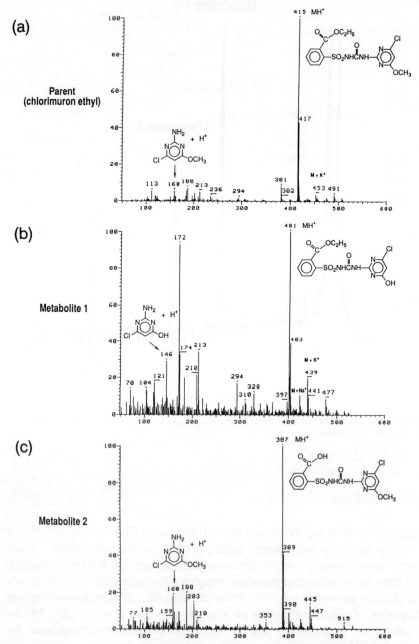

Figure 8.19. CF-FAB mass spectra of chlorimuron ethyl and microbial metabolites from an LC/MS analysis.

Acknowledgements

We thank W. R. Johnson and D. A. Suchanec for their skillful laboratory work.

REFERENCES

1. J.A. Romesser, and D.P. O'Keefe, *Biochem. Biophys. Res. Comm.* 140, 650 (1986).
2. A.C. Barefoot, and R.W. Reiser, *J. Chromatogr.* 398, 217 (1987).
3. A.C. Barefoot, and R.W. Reiser, *Biomed. Environ. Mass Spectrom.* 18, 77 (1989).
4. A.C. Barefoot, R.W. Reiser, and S.A. Cousins, *J. Chromatogr.* 474, 39 (1989).
5. P. Doberstein, E. Korte, G. Meyerhoff, and R. Pesh, *Int. J. Mass Spectrom. Ion Phys.* 46, 185 (1983).
6. J.G. Stroh, J.C. Cook, R.M. Millberg, L. Brayton, T. Kihara, Z. Huang, and K.L. Rinehart, Jr, *Anal. Chem.* 57, 985 (1985).
7. R.M. Caprioli, T. Fan, and J.S. Cottrell, *Anal. Chem.* 58, 2949 (1986).
8. R.W. Reiser, *Proceedings of the 37th ASMS Conference*, 21-26 May 1989, Miami Beach, FL, p. 1010.

8.4 ANALYSIS OF METABOLITES OF FATTY ACID OXIDATION

Daniel L. Norwood and David S. Millington

The catabolic pathways for fatty acids in mitochondria involve acyl-coenzyme A (CoA) compounds which act as metabolic intermediates, and acylcarnitines which act as cross-membrane transport molecules. The acylcarnitines (Scheme 8.1) are quaternary ammonium salts, and acyl-CoA's (Scheme 8.2)

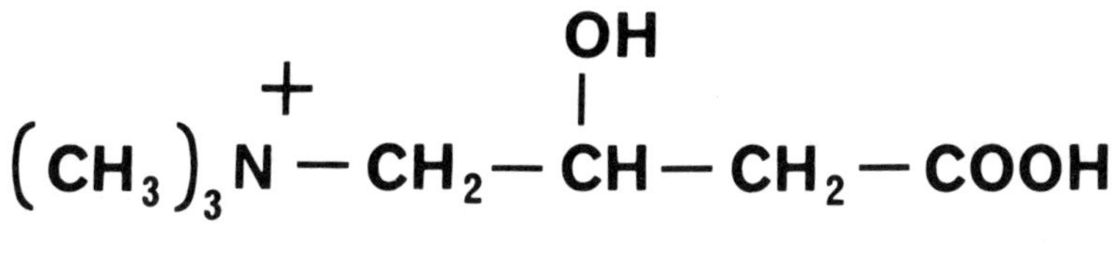

$$(CH_3)_3 \overset{+}{N} - CH_2 - \overset{OH}{\underset{|}{CH}} - CH_2 - COOH$$

Scheme 8.1.

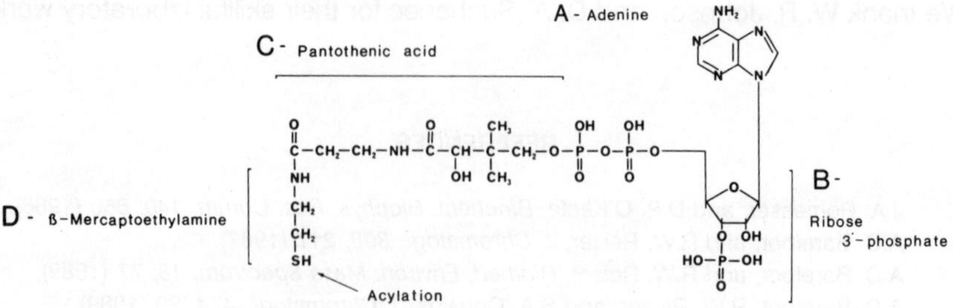

Scheme 8.2.

are complex molecules which include sub-units of (A) adenine, (B) ribose-3'-phosphate, (C) pantothenic acid, (D) beta-mercaptoethylamine, and the acyl group. Due to their lack of volatility and thermal instability, neither molecular type is amenable to gas phase ionization. Thermospray ionization (1) produces excessive fragmentation in acylcarnitines and usually no molecular weight information for acyl-CoA compounds.

This section describes the development and application of a continuous-flow (CF) FAB LC/MS system for the analysis of acylcarnitines and acyl-CoA compounds. The system employs different LC column types ranging from analytical (3.9 mm i.d.) to microbore (1 mm i.d.) and ultramicrobore (0.32 mm i.d.) with appropriate flow-splitting arrangements. Each system incorporates solvent gradients and mobile phase additives, with glycerol as a FAB matrix added to the mobile phase. Properties of the system such as sensitivity, robustness, chromatographic characteristics, and applicability to biological matrices are summarized.

8.4.1 Experimental

The CF-FAB LC/MS system is shown schematically in Figure 8.20. The system employed standard pulse-damped HPLC solvent delivery (Waters 600-MS) capable of delivering a linear binary solvent gradient. Injector 1 was either a Rheodyne 7125 with a 20 μl loop for the analytical column (3.9 mm i.d.), or a Valco CI4W with a 0.1 μl rotor for the microbore (1 mm i.d.) column. When the

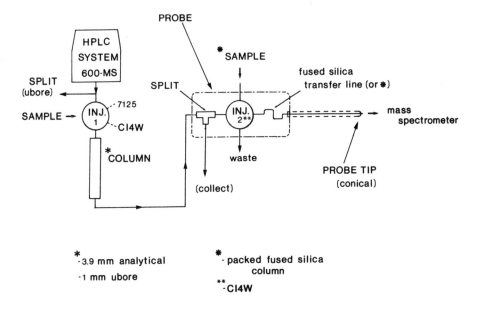

Figure 8.20. Schematic diagram of CF-FAB LC/MS system showing different column and injector configurations.

microbore column was in place, a splitter was incorporated between the pump and injector 1 to allow the pump to operate at 1 ml/min and deliver a 50 μl/min flow to the column. This arrangement allows gradient changes to reach the microbore column in a reasonable time period. A second splitter was utilized after the column to reduce the flow into the mass spectrometer from 1 ml/min (analytical column) or 50 μl/min (microbore column) to 10-15 μl/min. When the ultramicrobore 0.32 mm i.d. packed fused silica column (RoSil C18; Alltech Associates) was employed, a second CI4W injector was mounted in the CF-FAB probe handle (VG Masslab) and the column replaced the normal fused silica transfer line. Since this column operated at 10-15 μl/min, no effluent splitting into the mass spectrometer was required. The mass spectrometer employed was either a VG Trio-2 single quadrupole or VG Trio-3 triple quadrupole instrument (VG Masslab). Additional information regarding the design of the system, HPLC operating parameters, and mass spectrometer conditions for FAB have been given previously (2-4).

8.4.2 Brief of Applications

The CF-FAB LC/MS system with packed fused silica column has been employed for the analysis of short- and medium-chain (C_2-C_{10}) acylcarnitine standards, and diagnostic acylcarnitines derived from urine of children with inborn errors of fatty acid and branched-chain amino acid catabolism (2,3). Similar studies have been accomplished with an analytical column system (Waters µBondapak C18; 3.9 x 150 mm), an example of which from a patient with suspect medium-chain acylcoenzyme-A dehydrogenase (MCAD) deficiency is shown in Figure 8.21. Each system employed various binary

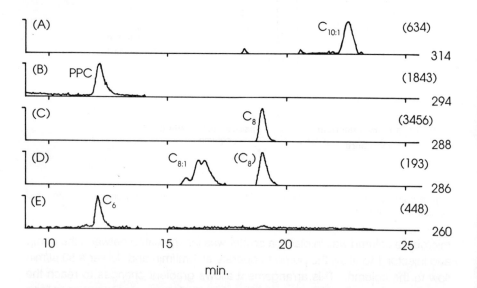

Figure 8.21. Separation and identification of diagnostic acylcarnitines from the urine of a patient with suspect medium-chain acylcoenzyme A dehydrogenase (MCAD) deficiency. Chromatograms are extracted ion current profiles of the molecular species (M^+) for various medium-chain acylcarnitines are shown in the figure (PPC is phenylpropionyl carnitine).

linear gradients of 0.05 M ammonium acetate (5% glycerol) and methanol (5% glycerol). The analytical column system was also shown to be applicable to long-chain (C_{14}-C_{18}) acylcarnitines and possessed sufficient sensitivity for most clinical applications.

Acyl-CoA compounds of various acyl group chain-length (C_2-C_{18}) were analyzed with either an analytical column (Waters NovaPak C18; 3.9 x 150 mm) or a microbore system (Keystone Scientific Hypersil BDS; 1.0 x 250 mm) employing binary linear gradients of ammonium acetate (2% glycerol) and acetonitrile (4). Analytes containing various functional groups within the acyl moiety (-C=C-; -OH; -COOH) were also analyzed. Although the acyl-CoA compounds fragment extensively in FAB, molecular weight information, i.e. $(M+H)^+$, was obtained for all examples investigated. The microbore system has been applied to the study of an enzyme catalyzed reaction (crotonase) to follow disappearance of the substrate crotonyl-CoA and appearance of product 3-hydroxybutyryl-CoA, as shown in Figure 8.22.

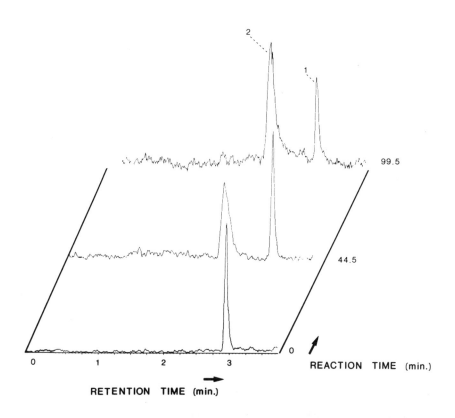

Figure 8.22. Combined extracted $(M+H)^+$ ion current profiles obtained at three time points during the crotonase catalyzed hydration of crotonyl-CoA (peak 1) to produce 3-hydroxybutyryl-CoA (peak 2). Isocratic CF-FAB LC/MS was employed with the 1 mm (i.d.) microbore column (reference 3).

8.4.3 Summary of System Performance

Conclusions and observations regarding the performance of the CF-FAB LC/MS system with its various operating modes may be summarized as follows:

(a) No significant chromatographic band broadening was observed with any column configuration employed for either acylcarnitines or acyl-CoA compounds.
(b) Sensitivity for both compound classes was sufficient for biological and clinical applications.
(c) Optimum system performance and stability were achieved with flow rates to the probe tip of 10-15 µl/min, source temperature of 45-50 °C, and saddle-field ion gun (Xe) voltages of 5-6 keV.
(d) Approximately 4-6 hours of stable operation could routinely be achieved with any system configuration regardless of mobile phase composition.
(e) No irregularities of flow along the fused silica transfer line were observed with any system configuration.

8.4.4 Conclusion

At this time, CF-FAB LC/MS is the most widely applicable analytical technique to metabolites of fatty acid catabolism.

REFERENCES

1. D.S. Millington, in *Mass Spectrometry in Biomedical Research,* edited by S.J. Gaskell, John Wiley, Chichester, 1986.
2. D.L. Norwood, N. Kodo, and D.S. Millington, *Rapid Commun. Mass Spectrom.* 2, 269-272 (1988).
3. D.S. Millington, D.L. Norwood, N. Kodo, C.R. Roe, and F. Inoue, *Anal. Biochem. 180,* 331-339 (1989).
4. D.L. Norwood, C.A. Bus, and D.S. Millington, *J. Chrom. (Biomed. Appl.).* In press (1990).

8.5 APPLICATIONS OF COAXIAL CF-FAB WITH TANDEM MASS SPECTROMETRY

Leesa J. Deterding and M. Arthur Moseley

Coaxial continuous-flow fast atom bombardment (CF-FAB) has proven to be a useful technique for interfacing nanoscale capillary liquid chromatography and capillary zone electrophoresis (CZE) with mass spectrometry (1-6). The coaxial CF-FAB interface, which we have recently developed and is described in detail in Chapters 5 and 6 of this volume, consists of a fused-silica capillary column, typically 10-15 μm (i.d.) 150 μm o.d., or a nanoscale packed microcapillary column, typically 50 μm (i.d.), 150 μm (o.d.) which is surrounded by a second fused-silica capillary column 200 μm (i.d.), 350 μm (o.d.) through which the matrix is introduced. These coaxial columns are interfaced using a 1/16 inch stainless steel tee (1,2). In this manner, there is no mixing of the matrix with the LC analytes until both have reached the FAB target tip, and the analyte flow rate and matrix flow rate and composition can be independently optimized. In addition, this coaxial design is advantageous since these nanoscale capillary LC columns offer much greater separation efficiencies (e.g., greater than 1×10^6 theoretical plates than conventional columns (7)).

We have demonstrated the ability to obtain on-the-fly mass spectra from low femtomole levels of a variety of compounds while maintaining high separation efficiencies (1-4). The coaxial CF-FAB interface provides great sensitivity due to the ability of the nanoscale capillary LC system to deliver a high concentration of analyte in a short time period. This advantage, combined with the ability to introduce a constant flow of analyte into the mass spectrometer, makes this interface very useful for the acquisition of MS/MS spectra (5).

8.5.1 Nanoscale Capillary LC-Coaxial CF-FAB/MS/MS

The coaxial CF-FAB interface is designed such that MS/MS spectra can be acquired from a constant flow of analyte into the mass spectrometer as well as from short injection time periods. The constant flux of analyte provides a steady production of ions for relatively long periods of time and can be advantageous for the acquisition of collisional activation decomposition (CAD) spectra, e.g. MS/MS/MS spectra, that are otherwise difficult to obtain on short-lived ion intensities (5). In addition, the need to continually reload the FAB probe tip is obviated. To make an injection onto the open-tube capillary column for an MS/MS experiment, approximately 200 μl of analyte solution is needed to fill the sample tee and tubing. Operating at the typical nanoscale capillary LC flow rate of <60 nl/min, this injection volume can deliver a constant flow of analyte for many hours.

The MS/MS spectra of a variety of compounds have been obtained with this technique (5). Figure 8.23 shows the CAD spectrum of 35 ng of bradykinin

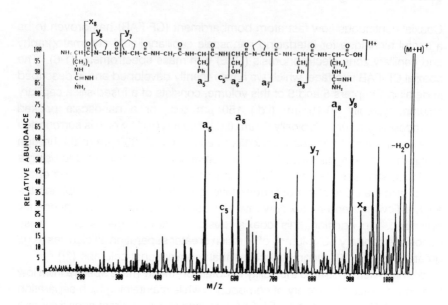

Figure 8.23. Coaxial CF-FAB/MS/MS spectrum of the $(M + H)^+$ ion of 35 ng of bradykinin acquired from a continuous flux of analyte. (Reprinted with permission from reference 5.)

which was acquired over a period of 43 seconds. A number of ions diagnostic of the sequence can be seen. If sample volume is limited (i.e., 10 μl or less), an alternative injection method which consists of a stainless steel pressure vessel (6) can be used to deliver a constant flux of analyte. With this method, nanoliter volumes can be used for the acquisition of MS/MS spectra.

The flow-injection of picogram quantities of samples yields very short elution times, i.e., 5 to 20 second wide peaks at half-height, that can impose stringent performance criteria on the mass spectrometer. Figure 8.24 demonstrates the ability of the coaxial CF-FAB interface to obtain MS/MS spectra on these short-lived samples. In this application, the mass spectrometer is set to scan the desired mass range at 3 s/decade which results in a total acquisition time of approximately 2 seconds. Because the peak width at half-height for the injections in Figure 8.24 is 6 seconds, two to three MS/MS scans can be obtained for each eluting peak. The MS/MS spectra of the $(M + H)^+$ ion of Met-Leu-Phe were acquired from a 220 pg and a 22 pg injection (Figures

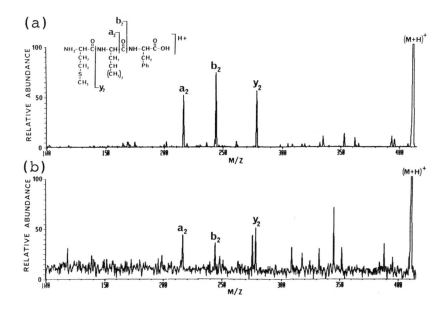

Figure 8.24. Coaxial CF-FAB/MS/MS spectra of (a) 220-pg and (b) 22-pg of the $(M + H)^+$ ion of Met-Leu-Phe acquired on-line. (Reprinted with permission from reference 5.)

8.24(a) and (b), respectively). The protonated molecular ion of m/z 410 fragments, giving ions of type a_2, b_2, and y_2.

The acquisition of MS/MS spectra using coaxial CF-FAB is also compatible with the short elution times involved in nanoscale capillary liquid chromatography. Using a nanoscale packed microcapillary column, the MS/MS spectra of the analytes of a mixture can be obtained as they are separated and elute from the column. The MS/MS spectrum of the $(M + H)^+$ ion of nanogram amounts of proctolin separated from a mixture is shown in Figure 8.25. Fragmentations resulting from typical backbone cleavages are observed.

8.5.2 CZE-Coaxial CF-FAB/MS/MS

The high separation efficiencies of capillary zone electrophoresis coupled with tandem mass spectrometry makes this a very powerful technique for the separation and identification of components of mixtures. The primary advantage of CZE over liquid chromatography is the superior separation efficiencies which can be obtained with CZE. However, because the peak

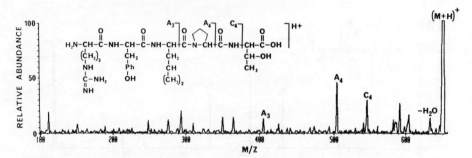

Figure 8.25. Coaxial CF-FAB/MS/MS spectrum of the (M + H)$^+$ ion of proctolin acquired from a separation of a mixture using a nanoscale packed microcapillary column.

widths of CZE peaks are very small, typically 1-5 seconds wide at half-height, the acquisition of on-line MS/MS data from a CZE separation can be difficult. In this case, one must compromise the separation efficiency somewhat by introducing a larger volume of sample to obtain good MS/MS spectra. The peaks are only slightly broadened and make the acquisition of MS/MS data more facile.

The CZE system (3) which is described in detail in Chapter 6 of this volume consists of a +30 kV power supply and a buffer-filled open tubular capillary which terminates at the +8 kV FAB probe tip. Therefore, a potential drop of 22 kV occurs across the length of the CZE column. The CZE buffer solution is typically 0.005 M ammonium acetate adjusted to pH 8.0 with ammonium hydroxide. Figure 8.26 shows the MS/MS spectrum of the (M + H)$^+$ ion of

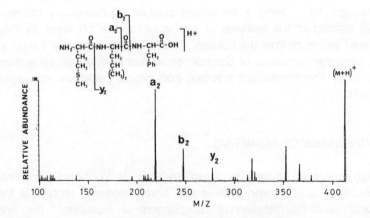

Figure 8.26. CZE/coaxial CF-FAB/MS/MS spectrum of 130 fmol of the (M + H)$^+$ ion of Met-Leu-Phe acquired on-line. (Reprinted with permission from reference 3.)

Met-Leu-Phe which was acquired from the electrophoretic separation of a mixture of Met-Leu-Phe and proctolin. The MS/MS spectrum was acquired on-line from approximately 130 fmol of the peptide.

REFERENCES

1. J.S.M. de Wit, L.J. Deterding, M.A. Moseley, K.B. Tomer, and J.W. Jorgenson, *Rapid Commun. Mass Spectrom. 2,* 100-104 (1988).
2. M.A. Moseley, L.J. Deterding, J.S.M. de Wit, K.B. Tomer, R.T. Kennedy, N. Bragg, and J.W. Jorgenson, *Anal. Chem. 62,* 1577-1584 (1989).
3. M.A. Moseley, L.J. Deterding, K.B. Tomer, and J.W. Jorgenson, *Rapid Commun. Mass Spectrom. 3,* 87-93 (1989).
4. M.A. Moseley, L.J. Deterding, K.B. Tomer, and J.W. Jorgenson, *J. Chromatogr. 480,* 197-209 (1989).
5. L.J. Deterding, M.A. Moseley, K.B. Tomer, and J.W. Jorgenson, *Anal. Chem. 61,* 2504-2511 (1989).
6. S. Pleasance, P. Thibault, A.M. Moseley, L.J. Deterding, K.B. Tomer, and J.W. Jorgenson, *J. Am. Soc. Mass Spectrom.* Submitted for publication.
7. J.W. Jorgenson, and E.J. Guthrie, *J. Chromatogr. 255,* 335-345 (1983).

INDEX

Acetogenins, LC/MS of   140-3
Acylcarnitines
  LC/MS of   175
  microbore LC/MS of   178
Apolipoprotein A1, microbore LC/MS of   97

Background
  and mass resolution   36
  at low mass   11
  effect on quantitation   50
  from glycerol   34
  reduction of   14
Batch samples
  analysis of   63
  processing   68-72
Benzo[a]pyrene, analysis of metabolites of   143-5
Bile conjugates   165
Biological applications   63-92
Biomedical applications   160-6
Biopharmaceuticals, LC/MS of   151

Capillary bore LC/MS   107-18
  biomedical applications   161
  of bile conjugates   165
  of carbohydrates   115
  of glycopeptide antibiotics   162
  of growth hormone   154
  of herbicide metabolites   166
  of peptide digests   152-3
Capillary zone electrophoresis, see CZE
Capillary zone electrophoresis/mass spectrometry, see CZE/MS
Carbohydrate analysis, LC/MS   115

Cesium ion gun
  focused   56-60
  unfocused   48
CF-FAB
  advantages   11-25, 29-30
  as LC/MS interface   94
  disadvantages   25-6
  high mass   25
  of non-polar compounds   137
  on MAT90   3
  on MS25   138
  on MS50   2, 138
  on TSQ70   6-8, 46-7
  on ZAB SEQ   31-3
  operational stability   4-11
  probe design   137-40
  sensitivity   14
CF-FAB/tandem MS   38-40
CF-LSIMS, instrumentation   45-8
Chlorosulfurons, LC/MS   170
Co-axial CF/FAB/tandem mass spectrometry   181-5
Co-axial interface
  for CZE/MS   126-30
  for LC/MS   107-10
  sensitivity   114
  with MS/MS   181-5
Conjugates
  glucuronic acid   144
  sulfate   145
Constant flow, on-line   79-80
Continuous-flow/liquid secondary ion mass spectrometry (LSIMS), see CF-LSIMS
Continuous-flow FAB, see CF-FAB

INDEX

CZE
  electromigration in 123
  electro-osmosis in 123
  flow profile 123
  sample injection 124
CZE/MS 121-36
  advantages 135
  apparatus 121
  buffers 129, 132
  co-axial interface 126-30
  combination interface 131
  limitations 135
  liquid juncture interface 125-6
  of peptides 130
    with MS/MS 183-5
  of tryptic digest 133-4
  sample injection 130

Displacement chromatography 151
  of growth hormone 156
Distearins 147-8

Electromigration in CZE 123
Electro-osmosis in CZE 123
Enzyme mixtures
  on-line 76
  peptidases 76

FAB MS 1
Fast atom bombardment mass spectrometry (FAB MS), *see* FAB MS
Fatty acid oxidation metabolites 175-80
Filter pads for removing liquid 3, 54
Fittings, low dead volume 68
Flow injection 12-13, 64-8
  advantages 65
  batch sample processing 68-72
  linearity 13
  on-line 73-9
Frit-FAB
  as LC/MS interface 93
  probe design 153
  with protein digest 107

Glucuronic acid conjugates 144, 145
Glycerol
  background from 34
  ions from 2
Glycopeptide antibiotic mixtures 162

Growth hormone
  capillary bore LC/MS of 154
  displacement chromatography of 156
  microbore LC/MS of 103
  tryptic digest of 154

Herbicide metabolites 166-75
  capillary bore LC/MS of 166
  LC/MS of 166

In vivo drug monitoring 85-8
Interface for liquid sample introduction 67
Ion suppression 51
  decrease of 17-25
  in tryptic digest 20

LC/MS 93-120
  and chromatographic resolution 97
  biomedical applications 160-6
  carbohydrate analysis 115
  flow restrictors 112
  in MS/MS 115
  mass summary of 100-2
  microinjection 111-12
  non-aqueous solvents 148
  of acetogenins 143
  of acylcarnitines 175
  of biopharmaceuticals 151
  of chlorosulfurons 170
  of glucuronic and sulfate conjugates 145
  of herbicide metabolites 166
  phase-system switching 117
  pressure injection 111
  splitless/split injection 110, 112
  with frit-FAB 153
Leucine-enkephalin, trace analysis 38-40
Liquid chromatography/mass spectrometry, *see* LC/MS
Liquid juncture interface in CZE/MS 125-6
Liquid secondary ion mass spectrometry (LSIMS) 45

Mass resolution and background 36
Mass summary of LC/MS 100-2
Matrix
  advantages 1

INDEX

disadvantages 2
for CZE/MS 129
for non-polar compounds 141
Memory effect 68
Microbore LC/MS
  column sizes 96
  gradient elution 96
  in protein sequence verification 103
  instrumental set-up 95-6
  of acylcarnitines 178
  of apolipoprotein A1 97
  of growth hormone 103
  of trace peptide contaminants 103
  of tryptic digest 97, 103
  sensitivity of tryptic digest 99
Microdialysis 80-9
  absolute recovery 83
  in vivo drug monitoring 85-8
  membrane binding effect 85
  relative recovery 83
  response time 83
  temperature effect 84-5
  use with peptidases 88-9
MS/MS
  and quantitation 50
  and trace analysis 38-40
  co-axial interface with 181-5
  general operation 70
  in LC/MS 115
  in on-line drug monitoring 86-8
  of peptides, with co-axial interface 182
  with CZE 183-5

Non-polar compounds
  analysis of acetogenins 140-3
  analysis of benzo[a]pyrene metabolites 143-5
  analysis of distearins 147-8
  analysis of t-butyl phosphines 148
  applications 137-49

On-line analysis
  drug monitoring 85-8
  of enzyme digests 80
On-line monitoring, reaction analysis 73-80

Peptidases
  and microdialysis 88-9
  on-line 76, 80

used in mixtures 76
Peptide digests, capillary bore LC/MS of 152-3
Peptide map 102
Peptide sequencing, and enzymes 80
Peptides
  CZE/MS 130
    with MS/MS 183-5
  MS/MS, with co-axial interface 182
  trace contaminants 103
Phase system switching in target compound analysis 117
Platelet-activating factor (PAF) 40-2
Protein sequence verification 103

Quantitation 45-61
  and MS/MS 50
  and stability 13
  parameters affecting 48-55
  with stable isotopes 48

Sensitivity, compared to standard FAB 15-17
Stability 4-11
  achieving thin film 6-8
  and quantitation 13
  effect of backflow 4-5
  flow rate effect 8
  of FAB gun 9
  operating definition 9-11
  temperature effect 8, 33
Sulfate conjugates 145

Target materials 4
Trace analysis 29-43
  of leucine-enkephalin 38-40
  of platelet activating factor 40-2
Trypsin
  kinetic analysis 73
  with multiple substrates 73
Tryptic digest
  and time domain 78
  CZE/MS 133-4
  ion suppression in 20
  LC/MS analysis 97, 103
  of growth hormone 154
  sensitivity by LC/MS 99
Tryptic map, sequence order of 78

Urine, direct analysis 71